H. E Webster

On the Annelida Chaetopoda of the Virginian coast

H. E Webster

On the Annelida Chaetopoda of the Virginian coast

ISBN/EAN: 9783743418868

Manufactured in Europe, USA, Canada, Australia, Japa

Cover: Foto ©berggeist007 / pixelio.de

Manufactured and distributed by brebook publishing software (www.brebook.com)

H. E Webster

On the Annelida Chaetopoda of the Virginian coast

ON THE

Annelida Chætopoda

OF THE

VIRGINIAN COAST,

BY

H. E. WEBSTER.

In Advance of Vol. ix of Transactions of the Albany Institute.

ANNELIDA CHÆTOPODA OF THE VIRGINIAN COAST.

By H. E. WEBSTER.

[Read before the Albany Institute, Oct. 15, 1878.]

The Annelida catalogued and described in the following pages were collected in the summer months of 1874 and 1876, by the zoölogical expeditions which, for some years past, Union College has sent out during the summer vacation. The locality was in Northampton Co., Virginia, (Eastern shore of Va.), between the main-land and the line of outside islands. Collecting on the eastern shore is in many respects unpleasant. The coast is monotonous; there is very little variety of station, unless a change from soft black mud to softer blacker mud can be called variety. At low-water a great area is exposed, but from high-water mark to the edges of the channels it is always mud; and when the dredge is let down it comes up filled with the same variety of mud; of course under such circumstances the work itself can not be pleasant. However, there was abundance of life. At low water the flats were black with *Ilyanassa obsoleta* Stimp., and two species of Gelasimus were present in numbers that defied computation; oysters and blue-crabs were everywhere; *Amphitrite ornata* was so common that in many places their extended tentacles almost touched each other; *Marphysa sanguinea* appeared at every turn of the spade or haul of the dredge; *Nereis limbata, Drilonereis longa, Cirratulus grandis, Enoplobranchus sanguineus* and other worms were present in the mud in great numbers; small annelids and molluscs abounded among the oysters. By far the greater part of our work was done with the spade at low-water. With the exception of the Syllidæ and some other small forms, nearly every species dredged was also found between tides. In a few places we found what our boatmen called " rocky bottoms." The rocks were a thin layer of dead shells, that had been washed into the deeper parts of the channels and remained there. These shells had been very thoroughly excavated by a species of sponge and other boring animals, and in the galleries thus formed most of the smaller species of annelids were found.

The results of the work, so far as concerns the annelids may be summarized as follows:

Number of Families, represented, - - - 23
" Genera, - - - - - 49
" Species, - - - - - - 59

The number of families would by many be regarded as too small, as I have followed Grube and Ehlers, rather than Kinberg and Malmgren in regard to family limitations; using Eunicidæ, for example, to include Marphysa, Lumbriconereis and Staurocephalus, each of which has been referred (and perhaps properly) to a distinct family. In the generic classification, on the other hand, I have usually followed what may be called the modern arrangement. Nevertheless it seems very probable that the views of Prof. Grube as to the proper limitations of the genera of setigerous annelids are correct, and will ultimately prevail. Of the genera adopted, four are new and six have not previously been reported from our coast. Twenty-seven of the species are believed to be new, besides four previously described, but new to our coast.

I am under obligations to Prof. Verrill both for advice and for the use of specimens.

My thanks are also due to Mr. J. A. Lintner of the N. Y. State Museum of Natural History, who has used his wide knowledge and experience to supplement my deficiences both in knowledge and experience, in the kindest and pleasantest manner possible.

T. R. Featherstonhaugh, M. D., of Schenectady, N. Y.,

and Mr. Thomas McKechnie of Newark, N. J., rendered valuable assistance in collecting; digging and dredging with patience and even with cheerfulness which was all the more wonderful as they were not sustained by any deep affection for worms, crabs, molluscs, etc.

Fam. POLYNOIDÆ.

LEPIDONOTUS (*Leach*) *Knbg.*

LEPIDONOTUS SQUAMATUS *Knbg.*

PL. I, FIGS. 1-5.

Aphrodita squamata LINN. Syst. Nat., ed. x, p. 655.
Polynoë squamata SAV. Syst. des Ann., p. 22.
" " AUD. & M. ED. Littoral de la France, ii, p. 80, pl. i, figs. 10-16.
" " QUATR. Hist. Nat. des Ann., i, p. 218.
" " GRUBE. Fam. der Ann., p. 36.
Aphrodita punctata. Zool. Danica, iii, p. 25 (non figs. of pl. 96).
" " O. FAB. Fauna Groenlandica, p. 311.
Lepidonote punctata ŒRSTED. Ann. Dan. Consp., p. 12, figs. 2, 5, 39, 41, 47, 48.— Grön. Ann. Dors., p. 16.
Lepidonote armadillo LEIDY. Marine Invert. Faun. N. J. & R. I., Ex. Journ. Phila. Acad., series ii, vol. iii, p. 16, pl. xi, f. 54.
Lepidonotus squamatus KNBG. Eugenies Resa; Zoölogie ii. p. 13, pl. iv, f. 15.
" " JOHNSTON. Cat. British Worms, p. 109, pl. viii, f. 1,
" " MALMGREN. Nord. Hafs-Ann., p. 56.
" " Ann. Polych., p. 130.
" " BAIRD. Linn. Proc. Zoölogy, viii, p. 182.
" " VERRILL. Invert. Animals of Vineyard Sound, etc., in Report of U. S. Commissioner of Fish & Fisheries, Part I, p. 581, pl. x, figs. 40-41.
" " MÖBIUS. Untersuchung der Ostsee, p. 112.
Polynoë dasypus QUATR. Hist. Ann., i, p. 226.

L. squamatus of the American coast seems to differ somewhat from the European form. Comparing f. 1, pl 1, with Kinberg's figure (15, pl. 4, Eug. Resa), it will be seen that in our specimens the anterior eyes are larger; the lateral prolongations of the head, from which the lateral antennæ arise, much shorter; the peduncle of the median antennæ not so much depressed, etc. I have collected this species at many points from Maine to Virginia, and after careful comparison am satisfied that fig. 1, re-represents accurately the form of the head and appendages for our specimens. Not common.

LEPIDONOTUS SQUAMATUS *var*. ANGUSTUS.

Lepidonotus angustus VERR. Invert An. Vineyard Sound, etc., p. 581.

Prof. Verrill now regards the form described by him as *L. angustus* as a variety of *L. squamatus*.

Common on shells, etc., from low water to 10 fathoms.

LEPIDONOTUS VARIABILIS n. sp.

PL. I, FIGS. 6-11. PL. II, FIGS. 12-14.

Body narrow, of nearly uniform width throughout; slightly convex above and below. The head (pl. I, f. 6) is convex laterally, with a well marked depression running from the base of the middle antenna, nearly to the posterior margin. Eyes, lateral, circular; the anterior pair a little back of the centre, slightly larger than the posterior pair. Middle antenna about double the length of the lateral, nearly three times as long as the head, somewhat swollen at its outer third, then tapering rapidly to a filiform termination. Lateral antennæ with a slight enlargement midway, otherwise similar to middle antenna.

Palpi, a little shorter than middle antenna, swollen at base, diminishing rapidly and uniformly.

The tentacular cirri have elongated basal articles; the inferior are about the length of the lateral antennæ; the superior a little longer. In structure they are the same as the middle antenna.

Elytra, 12 pairs, completely covering the back. First pair nearly circular, the others oval, slightly broader behind than in front (pl. I, f. 7). Posterior and outer margin coarsely fringed; an isolated patch of fringe on the inner margin, not arising from the edge, but from the surface of the elytron near the edge. Along the inner border where there is no fringe, is a series of minute papillæ. When not magnified the elytra seem to be smooth. In reality they are covered with minute, sharp, spine-like projections. The color of these little spines is usually reddish brown, though they may be any shade of brown, or even white.

The dorsal cirri are about one-half the length of the middle antenna, and have the same form. They arise from a stout basal article, much swollen along its inner half; outer half cylindrical. They project by about one-third of their own length beyond the setæ. The feet are large, in length about equal to the width of the body, somewhat compressed, diameter suddenly diminishing at outer third, truncated externally. Setæ of dorsal ramus numerous, delicate, usually covered, with a single series of rather coarse denticulations (pl. II, fig. 12). Those of the lower ramus, stout, bidentate (pl. I, figs. 9, 10); superior tooth very little curved, projecting some distance beyond the inferior. Below the apex there are a few stout denticles. Of the terminal teeth, the lower is frequently worn off (pl. II, f. 13). The setæ of the lower ramus are somewhat variable in form, as shown in the figures. Those of the first segment are not bidentate (pl. I, f. 11). The ventral cirri arise at about the inner third of the feet, from a small, rounded, basal article; they are minute, conical (pl. I, f. 8). There are two very long anal cirri, longer than the middle antenna. The dorsal cirri of the last segment turn directly backward, and reach about one-half as far as the anal cirri, thus giving the appearance of four anal cirri. Color, variable. The head may be clear white, or with dark brown specks. The antennæ and all superior cirri are white, with a black or dark brown band on the enlarged portion. The palpi are usually dark brown at base, becoming lighter externally, with white tips; or they may be white throughout. The general color of the body above is some shade of brown, but the elytra vary much in their markings. Often there is a large whitish spot, with numerous minute brown spots, or the white spot may disappear. The brown sometimes becomes nearly black. In a few specimens the black occurs in large blotches; or the space usually occupied by the white spot may be black. The body beneath the elytra is yellowish white. The anal segment is brown or black, and the two or three segments preceding the anal have transverse markings of the same color as the anal segment. Ventral surface more nearly uniform in its markings than the dorsal. General color, yellowish-white. Margins of each segment with an irregularly shaped brown spot. A central, white or yellowish-white, line, on each side of which on each segment is a brown spot. The anal cirri are usually dark colored throughout, and the external enlargement is hardly perceptible.

On pl. II, f. 14, a head is figured which will be seen to differ much from f. 6; but it is only in the head that the two forms differ. In setæ, shape of feet, form and character of elytra, etc., they agree perfectly. At least half of the specimens corresponding to fig 14 were females. I could not determine the sex of the others. Possibly this may be a sexual form, though I have been unable to find any notice of sexual variations in this group of annelids. It will be seen that the anterior margin of the first segment is prolonged on each side of the middle line into little triangular folds, which in contracted specimens encroach on the posterior margin of the head. This is also the case with *L. variabilis*.

Length 10 to 20^{mm}; diameter 2 to 4^{mm}—some of the widest specimens not being more than 15^{mm} in length, while some of the longest are not over 3^{mm} in width. Number of segments (setigerous), 25.

ANTINOË *Knbg.*

ANTINOË PARASITICA *n. sp.*

PL. II, FIGS. 15-22.

A minute form, found under the elytra of *Lepidametria commensalis*. I have never seen a perfect specimen. Of the three found, two were without elytra; the third had a few remaining. Number of setigerous segments, 21.

Head (pl. II, fig. 15), divided into lobes in front by a deep triangular depression which is continued backward to the posterior margin, as a narrow, but well defined impressed line. The widest part of the head is about the middle third, where it is convex at the sides; but the lateral margins both in front and back of this part are slightly concave; dorsal surface convex.

Anterior eyes, situated on the middle line, widely separated, black, crescentic: posterior eyes small, circular, about half-way between the front pair and the posterior margin of the head. Middle antenna lost, except basal article: this last occupies in part the space between the lobes of the head, is large, but not very long. The lateral antennæ are without basal articles; they arise beneath the anterior margin of the head, and are smooth, conical, slightly compressed at base: their length is about half that of the head. Palpi, nearly twice the length of the head, smooth, cylindrical for their inner three-fourths, then suddenly decreasing in diameter.

Tentacular cirri arise from an elongated cylindrical base; superior lost; inferior as long as the head, inner half cylindrical, outer half conical; covered from near their origin for two-thirds of their length with rather coarse papillæ.

Elytra, probably 12 pairs; very minute, transparent, smooth, edge without appendages. They were colorless, with numerous very light brown specks and streaks. They barely covered the base of the feet, not reaching to the middle line of the body. The shape varies, some being as shown in pl. II, f. 16, others as in f. 17.

Dorsal cirri long, projecting beyond the setæ, in all respects similar to the tentacular cirri. Feet long, nearly cylindrical, bilabiate. Dorsal ramus a mere papilla on the upper surface of the foot. Setæ of dorsal ramus (pl. II, f. 18) delicate, one edge very finely denticulated, terminating in a long capillary point. Setæ of ventral ramus of several kinds: first and second setigerous segments, the setæ terminate in a single sharp point, at some distance below which is a long series of sharp teeth (pl. II, f. 20); on the remaining segments, except the last two, the setæ are bidentate, serrate below the apex, the teeth being arranged in short series. The superior terminal tooth is much curved (pl. II, f. 19). In the penultimate segment these setæ are replaced in part by strong hooked setæ; while in the last segment the hooked setæ are the only ones found (pl. II, figs. 21, 22). It seems probable that the function of these hooks is to hold on to the body of *Lepidametria commensalis*, under the elytra of which they are found. Length about 2^{mm}.

LEPIDAMETRIA n. gen.

Lateral margins of the head prolonged to form the bases of the lateral antennæ. Middle antenna with distinct basal article. A small facial tubercle. Body elongated, flattened.

Elytra numerous, completely covering the body, or leaving a naked median space of variable width; along the posterior part of the body not always in pairs. Setæ of dorsal ramus, few, delicate : of the ventral ramus for the most part bidentate, but with a few, stouter than the others, pointed.

This genus is closely related to HALOSYDNA *Knbg.* but differs in the number and arrangement of the elytra, in having pointed setæ in the lower ramus, and in having a distinct facial tubercle. In many respects it agrees with LEPIDASTHENIA *Mgrn.*, but is excluded from that genus by having setæ in the dorsal rami.

LEPIDAMETRIA COMMENSALIS n. sp.

PL. III, FIGS. 23-31.

Head (f. 23), convex, sides rounded, widest and highest at posterior third, posterior margin slightly concave.

Eyes circular, black, lateral, all small, and in alcoholic specimens seen with difficulty. Anterior pair largest, situated at the widest part of the head. Posterior pair mere specks, half way between the front pair and the hind margin of the head.

Lateral antennæ, bases forming about one-third the length of the entire head; antennæ a little longer than the head. Base of median antenna stout, cylindrical, projecting a little beyond the origin of the lateral antennæ. Middle antenna one-third longer than the lateral. All the antennæ, together with the tentacular cirri and dorsal cirri are cylindrical to near the end, where there is a very slight swelling, after which they taper rapidly, having a filiform termination.

Palpi, stout, extending beyond the middle antenna, decreasing regularly to near the end, where they become suddenly filiform; covered with minute cylindrical papillæ.

Tentacular cirri arise from a long basal article, much enlarged at origin. The superior cirrus is a little shorter than the median antenna; inferior, a little shorter than the superior. The elytra are smooth, border without appendages. There may be from 38 to 50 on a side. They can not be enumerated in pairs, since opposite sides of the same segment may bear, one, an elytron, the other, a dorsal cirrus. For the first 32 segments the arrangement is uniform; viz: 1, 3, 4, 6,——26, 27, 29, 30, 32. After the 32d segment no two specimens present exactly the same arrangement; even the opposite sides of the same specimen, as mentioned above, differing both in the number and position of the elytra. For example, on one specimen between the 39th and 44th segments, inclusive, was the following arrangement. Right side, elytra on 39th and 44th segments:
Dorsal cirri, 40, 41, 42, 43.
Left side, elytra, 39, 41, 42, 44:
Dorsal cirri, 40, 43.

The elytra (figs. 25, 26, 27) extend along the entire length of the body, and in some cases cover the body completely, but usually there is a naked median space of variable width, and often they do not overlap, or even touch each other in any direction. Anterior pair of elytra circular, elsewhere oval, longer axis transverse, anterior margin emarginate when overlapped by preceding elytron; otherwise rounded. A variable number of posterior segments are without elytra. These segments are very short, and are always covered by the last pair.

The dorsal cirri arise from stout, nearly cylindrical basal articles (f. 28), which are one-third as long as the cirri. They extend a little beyond the foot. Their structure is the same as that of the antennæ, save that the subterminal swelling is even less obvious. The basal article originates a little outside of and behind the attachment of the elytra. Ventral ramus of foot stout, elongate, conical, widely excavated for the transmission of the setæ, and obliquely truncated from above downward. Dorsal ramus a mere papilla. Setæ of dorsal ramus (f. 31) few, delicate, tapering gradually to a minute capillary ending, one edge finely denticulated. One stout concealed acicula.

Ventral setæ in two bundles; superior with one or two stout setæ (f. 29) usually with bluntly rounded apex, beneath which for a short distance one edge is minutely denticulated. In the upper bundle there are also from one to three bidentate setæ. Below the apex on each side is a series of rather long sharp teeth, from 6 to 8 in number. These setæ are not quite so large as those mentioned above. In the lower bundle are from 6 to 12 bidentate setæ every way similar to those of the upper, but a little more delicate (f. 30). All the setæ however are stout and long, except those of the anterior feet which are delicate.

The first ventral cirrus is in all respects similar to the dorsal cirri; its basal article is long, and arises from the base of the foot. After this one they originate at the inner third of the foot from a small papilla. They are minute, conical. At the outer posterior angle of each segment, below, is a small cylindrical papilla, projecting backward and downward, not visible on a few of the anterior segments.

Facial papilla. Just in front of the mouth, beneath the base of the median antenna is a small, facial tubercle or papilla, bluntly rounded externally. Anal cirri two, similar to dorsal cirri. Body very slightly convex above and below.

Color. When divested of the elytra the dorsal surface is usually reddish-purple, with narrow bands of darker purple between the segments. Some specimens again are dark gray, others nearly black. The elytra are transparent, but when removed from the body show a light brown or gray background with numerous dark brown or black spots and blotches. While the elytra remain on the body their circular attachments show through as a series of well marked white spots. Head and facial papilla purple. Antennæ black at base, succeeded by white with brown markings, then comes a black ring corresponding to the enlarged portion; terminal part white. Palpi white. Tentacular cirri, first ventral cirrus, dorsal cirri, and anal cirri colored same as antennæ; the last dorsal cirri and the anal cirri are sometimes darker, or nearly black throughout. Ventral cirri white, flecked with brown at base. Ventral surface same as dorsal. Dorsal setæ, and bidentate setæ of ventral ramus, amber-yellow; single pointed large setæ, dark reddish-brown.

Besides the specimens found in Virginia, one specimen has been found by Prof. Verrill near New Haven, Conn.

Length, 50–90^{mm}.

Breadth, including feet, 5–7^{mm}.

Number of segments, 60–80.

Lives in tube of *Amphitrite ornata* Verrill.

Fam. SIGALIONIDÆ.

STHENELAIS *Knbg.*

STHENELAIS PICTA *Verrill.*

VERR. Invert. Animals of Vineyard Sound, etc., p. 582.

Fam. NEPHTHYDIDÆ.

NEPHTHYS *Cuv.*

NEPHTHYS INGENS *Stimpson.*

STIMP. Mar. Invert. of Grand Manan, p. 33, in Smithson. Contrib., 1854.
VERRILL. Invert. An. Vin. Sound, etc., p. 583, pl. xii, figs. 59, 60.

NEPHTHYS PICTA *Ehlers.*

EHL. Borstenwurmer, p. 632, pl. xxiii, figs. 9, 35.
VERR. Invert. An. Vin. Sound, etc., p. 583, pl. xii, f. 57.

Rare in Va.

Fam. PHYLLODOCIDÆ.

PHYLLODICE (*Sav.*) *Mgrn.*

PHYLLODICE FRAGILIS n. sp.

PL. III, FIGS. 32-37.

Body elongated, flattened, widest in the middle, tapering uniformly in both directions, first and last segments very narrow; middle third above, slightly convex; elsewhere, both above and below, flattened.

Head (f. 32) wider than long, very convex, bluntly rounded in front, posterior angles rounded, posterior margin very slightly convex, lateral margins convex, widest at posterior third. Eyes large, circular, black, situated at the widest part of the head, nearly lateral.

Antennæ, inferior slightly longer than superior, all fusiform, stout, not as long as the head. The first segment is longer than the second; it bears three pairs of tentacular cirri. The first pair originate close to the lateral margin of the head, the second, back of and a little within the first, the third, under the second. The second segment bears one pair of cirri (the 4th). The second pair are the longest, reaching back to the fifth segment. The others are about half as long. They all arise from short cylindrical basal articles. Diameter nearly uniform to near the end, where they taper slightly to a bluntly rounded termination.

The incisions for the feet extend inward so far that the feet on each side form one-third of the width of the body.

The feet are crowded, a short free portion is cylindrical, bluntly rounded externally, bilabiate, transmitting a fan of delicate setæ.

The stem of the setæ is much longer than the appendix somewhat flattened near the end, and with a transverse terminal series of minute teeth. The appendix is short, usually curved, wide at base, rapidly diminishing (f. 36). The appendix however may be double the length of the one figured, and less curved.

The dorsal cirri (branchiæ) are somewhat variable in form: but usually broad heart-shaped on the anterior segments (f. 33), becoming somewhat narrower, and less bluntly rounded externally further back (f. 34), while on young specimens, and on recently renewed lost parts, they are narrower, becoming somewhat lanceolate (f. 35). The ventral cirri (f. 37) extend beyond the feet, are nearly cylindrical, bluntly rounded externally.

Anal cirri shaped much like antennæ, may be as long as, or double, the length of the antennæ.

Color. Head greenish-white. Body from light to dark green. All cirri generally yellowish-green. Body and dorsal cirri with numerous, irregularly placed, dark brown spots. On young specimens and on renewed parts the color is always light, and the brown spots few, or absent.

Length of longest specimen 50^{mm}. Greatest diameter of same specimen 2^{mm}.

Found on shells between tides. Occasionally dredged. Common.

EUMIDA *Mgrn*.
EUMIDA MACULOSA *n. sp.*
PL. IV, FIGS. 38-41.

Head (f. 38) convex, rounded in front, posterior emargination distinct but not deep. Anterior antennæ in length about equal to the head, rather stout. Posterior (unpaired), a little shorter, and more delicate than anterior. Eyes posterior, lateral, circular, large, black. Of the tentacular cirri those of the first and second pairs are about half as long as the third; the third (anterior superior), reaches back to the 8th or 9th seg; the 4th about two-thirds as long as the 3d. The first segment is a little longer than the 2d, about equal to the 5th.

Dorsal cirri, on the anterior segments small, lanceolate (f. 39), becoming broader and heart-shaped (f. 40), while on the posterior segments they become again elongate, pointed, but with basal two-thirds of uniform width.

Ventral cirri small, lanceolate, sometimes with one straight and one convex side, reaching slightly beyond the feet; somewhat longer on posterior segments than on anterior.

Anal cirri about the length of the last four segments, arising from a stout basal article forming about one-third their entire length. Terminal portion, subulate, acute. The extended proboscis is about one-third as long as the body. It increases regularly in diameter from its origin, ending in a crown of papillæ. When highly magnified, longitudinal rows of cylindrical papillæ can be seen.

Color of the body, yellowish-white with numerous minute gray spots; appendages white. Dorsal cirri tinged with green.

On shells; 5 to 10 fathoms.

The specimen from which figs. 38, 39, and 40 were made was 3.5^{mm} long. Another specimen was 10^{mm} in length; diameter, 1^{mm}. Number of segments, 65.

Fam. HESIONIDÆ.

PODARKE *Ehlers*.

PODARKE OBSCURA *Verrill*.

VERR. Invert. An. Vin. Sound, etc., p. 589; pl. xii, fig. 61.

Six to twelve fathoms, on shells. Rare.

Fam. SYLLIDÆ.

SYLLIS (*Sav.*) *Ehlers.*

SYLLIS GRACILIS *Grube.*

GRUBE. Actinien, Echinodermen und Würmer, p. 77.
CLAPARÈDE. Glanures Zoötomiques parmi les Annélides de Port-Vendres, p. 75, pl. v, f. 3.
CLAPARÈDE. Annel. Chétopodes du G. de N., p. 503, pl. xv, f. 3.
MARION ET BOBRETZKY: in Ann. des Sciences Naturelles, 6th series, vol. xi, p. 23, pl. ii. f. 6. (1875.)

The specimens from Virginia agree perfectly with the descriptions given by Grube and Claparède. As remarked by Marion and Bobretzk, the figures of the furcate setæ given by Clpd. are very inaccurate. Those of M. and B. are very good, and removed the last doubt as to the identity of the Virginian specimens with *S. gracilis* Gr. They are, however, smaller than even those described from Port-Vendres, the largest specimen being only 11^{mm} long.

In some specimens the furcate setæ extend to the last segment, to the complete exclusion of the compound setæ; in others both furcate and compound setæ occur on a variable number of posterior segments, while on others again, the compound setæ entirely replace the furcate on a few of the posterior segments.

Common on Oysters at low water, and on dredged shells, from 4 to 12 fathoms.

SYLLIS FRAGILIS *n. sp.*

PL. IV, FIGS. 42, 43.

Head convex, anterior and posterior margins nearly straight lines, passing into the rounded lateral margins. Width double the length.

Eyes six. Anterior pair mere specks just outside the bases of lateral antennæ; middle pair at posterior third of head, lateral, large; posterior pair a little smaller than the middle, back of and within the latter, but almost touching them; all dark red. Median antenna, arising between median pair of eyes, is always very long, but varies, being from 5 to 8 times the width of the body. Lateral antennæ about two-thirds as long as the median. Superior tentacular cirrus as long as median antenna; inferior, length of lateral antennæ.

Palpi, portion projecting beyond the head slightly longer than the head. They are broad, flattened, inner two-thirds of uniform width, outer third with external margin curved inward, internal margin straight, end bluntly rounded. They are free and diverge for their outer two-thirds, united by a membrane along their posterior third.

The dorsal cirri arise from short stout basal articles (or elevations of the body). Their length varies from four to eight times the width of the body. There is a strong tendency to alternation between the shorter and longer cirri. Anal cirri two, as long as the last 6 or 8 segments taken together. The antennæ, tentacular cirri, dorsal cirri and anal cirri are alike in form and structure. They are uniform in diameter from end to end, are not composed of distinct articles but are wrinkled, and sometimes so regularly as to give the appearance of distinct annulation. All are covered with short stiff hairs.

Ventral cirri arise about half-way out on feet. Anterior not projecting beyond the feet, but the posterior often reaching slightly beyond. They are delicate, conical.

The œsophagus reaches to the 4th or 5th segment. Anterior end with a circle of small, flattened, triangular papillæ; one stout conical tooth. The stomach is about the same length as the œsophagus, reaching to the 9th or 10th segment. Back of the stomach, a pair of lateral glands.

The feet are uniramous, conical, bilabiate, in length equal to the width of the body. The setæ (f. 42) are numerous, arranged in a fan, the upper as long or a little longer than the foot, each lower one becoming a little shorter. All are long. There are one or two (f. 43) aciculæ in each foot, which are straight, taper uniformly, and end in a small enlargement (button). The last 6 or 8 segments taper slightly; otherwise the body is of uniform width.

The general color of the dorsum is yellowish-white; feet and sides, white; ventral surface same as dorsal. Between the segments runs a dirty or brownish-white line, and there is a similar line between the bases of the dorsal cirri on each segment. The ground color is interrupted by numerous dark brown spots or specks, for the most part arranged in transverse lines. These spots are dark and very numerous on the dorsal and lateral surfaces of the feet.

Sexual Forms.

The males and females of this species do not differ much from each other or from the stem form. I did not see many examples of either sexual form, and most of these were mutilated. So far as I was able to determine, the capillary (sexual) setæ in the male begin on the 11th or 12th segment, and fail on the last 8. They were short, not reaching beyond the feet. This part of the body was slightly swollen; color, pure white.

In the female, the capillary setæ begin on the same segment as in the male; they fail on the last 10 segments, are very long, double the length of the feet and ordinary setæ taken together. The eggs are brown, and determine the color of this part of the body. The anterior and posterior segments have the usual color of the stem form. On all sexual specimens, and on many that could not be determined as such, there is a dark brown crescent at the base of each foot (ventral surface).

Number of segments, in one specimen 38, in another 44.
Length, 4 to 5^{mm}.

ODONTOSYLLIS *Claparède*.

ODONTOSYLLIS? FULGURANS *Clpd.*

CLAPARÈDE. Glanures Zoötomiques, etc., p. 95, pl. viii, f. 1.
QUATREFAGES. Hist. des Ann., ii, p. 648.
MAR. ET BOBR : in Ann. des Sci. Nat., 6th series, vol. ii, p. 40, pl. iv, f. 11.

I am in doubt as to this form; in many respects it agrees perfectly with *O. fulgurans*, but in color, and in the structure of the feet and ventral cirri, it agrees better with *O. Dugesiana* (*Glanures Zoöt.*, p. 97, pl. 8, f. 2.)

On the whole, my specimens seem to confirm the supposition of MAR. and BOBR. (l. c. p. 41) that *O. fulgurans* and *O. Dugesiana* are not distinct species.

Sexual forms: male not found; female, two specimens. In one, the capillary setæ begin on the 21st setigerous segment, and exist on 14 segments, followed by 28 segments with compound setæ only; some few of the posterior segments lacking. In the other specimen the sexual setæ are on 19 segments, beginning with the 21st; rest of body lost. In both, the body back of the 20th segment was pure white, swollen with eggs, as far as the sexual setæ extended. On one specimen, the dark crescents described under *S. fragilis* were seen, in the other, not. In the second female mentioned above, the eggs were very large, evidently nearly mature, and the eyes on each side had coalesced.

Found on shells, stones, etc., dredged from 6 to 12 fathoms.

Length of a sexual form, from 4 to 12^{mm}.

Length of a nearly entire female, containing 64 segments, 12^{mm}.

SPHÆROSYLLIS *Clpd.*

SPHÆROSYLLIS FORTUITA *n. sp.*

PL. IV. FIGS. 44-48.

Head oval, slightly convex, length to breadth as 1 to 2 (f. 44). Eyes four, large, oval; anterior pair at posterior third of the head, lateral; posterior pair just within and but little back of the anterior.

Antennæ fusiform, equal, two-thirds as long as head and palpi taken together. The unpaired antenna originates between the posterior eyes; the lateral just in front of the anterior eyes. The palpi are large, rounded externally, bluntly rounded in front, united along their inner two-thirds, outer third free, but not diverging. There is a wide and deep depression between the palpi, extending from the head to the free portion.

First segment narrow, but perfectly distinct from the head.

Tentacular cirri similar in all respects to the antennæ.

Dorsal cirri, a trifle shorter than the antennæ, and not quite so much enlarged at base, especially on the posterior segments.

The anal cirri are double the length of the dorsal cirri, constricted at base; their diameter increases slightly along their inner half, when they are somewhat suddenly constricted, the outer half having about one-half the greatest diameter of the basal half (f. 45).

Ventral cirri, arise from base of feet, delicate, cylindrical, in front as long as the feet, longer behind and turned backward on a few of the posterior segments.

Feet very stout, cylindrical, bilabiate. The œsophagus reaches to the middle of the 5th segment. The stomach barely reaches into the 7th. The œsophagus has a covering of pigment, interrupted by a narrow clear band in the anterior part of the 4th segment.

Body convex above, segmentation well marked. A few of the anterior segments taper slightly. The body attains its greatest width at the 7th segment, remaining unchanged to the last 8 segments, whence it tapers quite rapidly. Anal segment narrow, without appendages save the anal cirri.

Small fusiform papillæ are scattered irregularly between the feet, and at their bases, much more numerously in the middle of the body than at either end.

Setæ, very short and of three kinds, two of which are compound. One form has a stout basal article (f. 47), a little longer than the appendix; there are 3 to 5 of these in each bundle. The second form (f. 46), longer than the first, has the appendix elongate, delicate, a little longer than the stem, one or two to each foot; in both, the edge of the appendix is beset with short, stiff hairs. Of the simple setæ (f..48) there is usually but one, sometimes two to each foot.

Length, 3. 5^{mm}; greatest diameter, 0.25^{mm}. Number of segments, 33.

A single specimen; not recognized when collected, but found afterwards in perfect condition in a lot of PÆDO-PHYLAX.

In many respects this species is similar to *S. pirifera* Clpd. (*Ann. de Naples*, p. 515, pl. xiv, f. 2), but differs from it in the shape of the head, in the free terminal part of the palpi, in the position of the eyes, and in the form of the antennæ and cirri.

I have referred this species to SPHÆROSYLLIS, though somewhat in doubt as to whether the name can be retained for it. Claparède's original diagnosis of the genus was defective and his figures misleading. Ehlers would seem to have been perfectly justified in ascribing to the genus five antennæ, no tentacular cirri, first segment with same appendages as the second: and in referring to the genus thus limited his new species *S. Claparèdii* (*Die Borstenwürmer*, p. 252, pl. ix, figs. 10-13). Claparède afterwards corrected his generic diagnosis, and gave a new figure of his type species *S. hystrix* (*Glanures*, p. 85, pl. 6, f. 1), but this was subsequent to the publication of Ehlers' work. In any case, if *S. hystrix* is to remain the type of the genus SPHÆROSYLLIS, then *S. Claparèdii* Ehlers, must be the type of a new genus.

PÆDOPHYLAX *Clpd.*

Annelides Chétopodes du Golfe de Naples p. 520.

Claparède draws one of his generic characteristics from the position of the eyes, assigning one pair to the buccal segment. This is not the case with the following species. In fact, it is not true even of *P. veruger* Clp., in which they are found in the depression between the head and buccal segment.

PÆDOPHYLAX DISPAR *n. sp.*

PL. IV, FIG. 49. PL. V, FIGS. 50-55.

Head (f. 49) convex, rounded in front and at the sides, posterior margin nearly straight, length to breadth as 1:2.

Eyes four, large, circular, lateral. Posterior pair a trifle smaller than the anterior, a little within, almost in contact with them, very near the posterior margin of the head.

Palpi: length of projecting portion nearly double the length of the head, convex, rounded laterally, diminishing forward, front bluntly rounded and slightly emarginate; above, the palpi are separated by a slightly impressed line; below, by a depression which widens and deepens behind.

Antennæ: median, arising near posterior margin of head, reaches nearly to the end of the palpi, fusiform;

lateral, arise just within anterior eyes, minute, mere papillæ, fusiform. Tentacular cirri, similar in all respects to lateral antennæ.

Dorsal and ventral cirri a little larger than the lateral antennæ; otherwise, the same.

Anal cirri two, slightly constricted at base; for the rest, subulate, in length about equal to last three segments, or to middle antenna. On one specimen there were three anal cirri, all of the same length.

Feet fleshy, swollen, in length about one-third the width of the body, a little longer than the dorsal cirri, bilabiate.

Setæ: on the anterior segments three or four with very short curved appendix (f. 51); one with capillary appendix (f. 52); one simple seta, slightly curved near the end (f. 54). On the posterior segments two or three of the first kind, one of the third, and one similar to the third, but more curved (f. 55).

Body of nearly uniform width in the anterior half; posterior half, tapering gradually, the last segment being about one-third the width of the middle segment. Number of segments, from 20 to 40. Length, 2 to 4^{mm}; diameter, 0.25 to 0.5^{mm}.

Color: either colorless, or white with a slight reddish tinge. Usually there is a white line between the segments.

Œsophagus, extends to the middle of the 5th segment. Stomach white, extending to the middle of the 8th segment.

Common on shells, stones, etc. Dredged, 4 to 10 fathoms.

AUTOLYTUS (Grube) Marenzeller.

Ehlers (*Borstenwürmer*) gives as a leading characteristic of AUTOLYTUS and PROCERÆA the failure of palpi. Claparède (*Glénures*) says of AUTOLYTUS, " Palpi (lobes frontaux) not projecting ; " and in his *Ann. du G. de N.* adopts Ehlers' diagnosis of PROCERÆA. Marion and Bobr. (*Ann. des Sci. Nat.*, Series vi, vol. 2) assign palpi to their *Autolytus* (*Proceræa*) *ornatus*. They regard PROCERÆA as a subgenus of AUTOLYTUS. Finally. Marenzeller (zur *Kentniss der adriatischen Anel.* aus dem lxxii Bande *der Sitzb. der K. Akad. der Wissensch.*) states that *Proceræa picta* Ehlers (type of the genus) has palpi, as well as two species described by himself (*P. luxurians* Mar., *P. macrophthalma* Mar.). He accordingly corrects (l. c. p. 37) Ehlers' diagnosis of the genus, and adds that the same is true for AUTOLYTUS. The following species will be seen to bear out this conclusion.

AUTOLYTUS HESPERIDUM *Claparède.*

CLPD. Annel. Chétopodes du G. de N., p. 526, pl. xiv, fig. 1.

My specimens agree with *A. hesperidum* Clpd. in every particular, save that they certainly possess projecting palpi. These form a thin rim, projecting a variable distance beyond the head, plainly divided below, separated above only by a shallow depression. The head proper is transversely oval, thicker and more convex than the projecting part. The outline of head and palpi together is exactly that given by Claparède. In the position of the central antenna given in his figure, the line between the palpi is of course concealed. This is often the case both in living and in alcoholic specimens. In fact, I examined a number of specimens without observing the palpi. When I did see them, I supposed that my previous identification was incorrect. It was, however, impossible to make a description, by which my specimens could be distinguished from *A. hesperidum*. I had not at that time seen the observations of M. and B., and of Marenzeller, cited above, by which my views have been confirmed. I can not find that *A. hespe-*

ridum has been examined by any one since it was described by Claparède. The extent of the projecting portion of the palpi is somewhat variable, but if in Claparède's figure (pl. 14, fig. 1), a curved line be drawn, passing just in front of the origin of the lateral antennæ, terminating on either side opposite the anterior (minute) eye on that side, the usual condition will be represented. Add, that though the antennæ often are as given by Claparède, in some specimens, the central antenna is longer than the lateral, sometimes as long as the first dorsal cirrus; also, that while both kinds of anal cirri (Clpd. l. c., pl. 14, f. 1 A) occur, I have never seen both on the same specimen.

Sexual Forms.

Female differs from the stem form in the following points: the head is very short, length to breadth as 1 to $2\frac{1}{2}$; middle third of anterior margin curving sharply backward, the antero-lateral portions appearing as rounded projecting lobes.

Antennæ, both median and lateral, arise from the anterior margin of the head; they are equal in length, being much shorter than those of the stem form.

Anterior eyes very large, posterior pair a little within the anterior, but nearly touching them. The anterior minute eye-specks of the stem form not seen.

The tentacular cirri, and dorsal cirri of the first segment, bear the same relation to the antennæ as in the stem form.

The capillary setæ are very long, beginning on the 4th, 5th, or 6th segment; not found on the last 4 to 7. Feet much swollen at base.

Eggs, very large, generally crowded, and then irregularly polygonal, otherwise spherical, their diameter about one-half the width of the body; of a dark purple color, with an eccentric clear nucleus.

Color of the body determined by the eggs, wherever they exist; elsewhere the same as in the stem form.

Number of segments, from 24 to 30. Length, 3 to 4^{mm}. Greatest diameter, 1^{mm}.

Specimens of the mature female occurred through July and August. They are quite delicate, and were very uneasy, swimming about constantly and rapidly, with a quick, undulatory motion. All the features mentioned above seem to originate after separation has taken place. I often found the stem form with eggs giving the characteristic purple color in the posterior segments, and with the separation carried so far that the slightest touch served to complete it, but in only one case did I find antennæ or tentacular cirri formed, or the peculiar outline of the head indicated, before separation. In this one specimen the antennæ and tentacular cirri were mere buds, the first segment was without feet; but the eyes and capillary setæ had not appeared.

Male. No adult male was found. The posterior segments of the stem often contained very fine granular matter, which I believe was the male element; it gave a very bright orange color to the segments in which it was found. No other change had taken place in these segments, though the division was often nearly complete, as mentioned above in the case of the female.

This species was common on shells, etc. Dredged, 4 to 12 fathoms.

PROCERÆA (*Ehlers*) *Marenzeller*.

EHLERS. Borstenwürmer, p. 256.
MARENZELLER. Zur Kentniss der adriatischen Anneliden, zweiter Beitrag, p. 37.

For remarks on this genus see under AUTOLYTUS.

PROCERÆA TARDIGRADA *n. sp.*

Head, quite thick, transversely oval, convex; head and palpi together nearly circular; palpi much the same as in *Autolytus hesperidum*; very thin, and the dividing line above not well marked, but existing.

Eyes small, circular, umber-brown, on the posterior part of the head, four in number; the anterior pair lateral, the posterior pair just within the anterior, very near the posterior margin of the head, smaller than the anterior.

Antennæ: the median arises between the anterior eyes, and when turned backward, reaches to the 8th segment; the lateral arise just in front of the anterior eyes, and are from one-third to one-half as long as the median.

Tentacular cirri: the superior about as long as the lateral antennæ; inferior one-half as long.

First dorsal cirrus, nearly as long as median antenna. Second dorsal cirrus, one-half as long as first.

The antennæ and tentacular cirri habitually coiled in a flat or a conical spiral. Antennæ, all superior cirri, and anterior margin of palpi, with numerous short, stiff hairs.

Dorsal cirri, a little less than width of body in length.

Anal cirri, two in number, a little longer than the longest dorsal cirri. The antennæ and all the cirri are irregularly wrinkled, and have nearly the same diameter from end to end.

The proboscis terminates in rounded fleshy lobes; the œsophagus, in a circle of flattened triangular papillæ. It extends to the 8th segment, doubling twice on itself just before entering the stomach. The stomach occupies from three to four segments.

Feet large, fleshy, cylindrical, bilabiate, anterior lobe slightly longer than posterior; incision between the lobes very shallow. They originate along the line of union of the lateral and ventral surfaces, and are directed downward.

Setæ arranged in a single series, and directed downward and backward; on the anterior segments about twelve to each foot, decreasing in number backward, till on the last segments there are only three or four. Body very convex above, flat below; width of the first fourteen segments uniform, after the 14th decreasing gradually to the minute terminal one.

The 1st, 2d and 3d segments are short; 4th and 5th equal in length and width; 6th to 14th, inclusive, length double the width; after the 14th, they become gradually shorter, the last ones being very short and crowded.

Color: general color of body yellowish-white. Antennæ, front of head and all cirri, white. Posterior part of head and first segment (in one case the second also), dark brownish-red. Between the eyes there is a narrow band, extending back to the middle of the first segment, of a darker shade than that of the head generally; and from the posterior eyes, on each side a similar line runs back to the third segment. Bands of the same color as the head cross the posterior part of the following segments (extending to the ventral surface, but not crossing it), viz: from the 3d to the 11th inclusive, 14th, 17th, 20th, and on every fourth segment after the 20th, except that the last three are on consecutive segments, and not well marked. On the anterior part of the body the width of these bands is about one-half the length of the 4th segment. As the segments shorten, after the 14th, this band also grows narrow. The lower surface of the base of the dorsal cirri, after the 7th, has a spot of the same color.

The movements of this species are peculiarly slow, the direction of its feet and setæ are such that it seems to *walk* on the tips of its setæ.

Length of specimen described, 11^{mm}; greatest diameter, 1^{mm}.

Number of segments, 80.

Length of head and first 14 segments, 5^{mm}.

On one specimen (lost before the examination was finished) a new head was forming back of the 14th segment.

Short antennæ had appeared, the lateral being bifurcate at outer third. The dorsal cirrus of the next segment had doubled its length.

On another specimen, there was a sudden falling off in width back of the 14th segment, while two other specimens had lost all save the first fourteen segments. It would seem that the separation of the sexual from the stem form takes place between the 14th and 15th segments.

Dredged : 4 to 12 fathoms; on shells, etc.

PROCERÆA? CŒRULEA n. sp.

The following notes were made upon two specimens, taken in August, both females, badly injured, and lost by accident before the examination was completed. They belong either to PROCERÆA or AUTOLYTUS; their mutilated condition did not admit of a positive generic reference.

Body, strongly convex above, flat below.

Eyes four; the anterior pair very large; the lateral, convex, nearly terminal; the posterior eyes, smaller, dorsal.

Antennæ : median very long, arising back of anterior eyes; lateral, half as long as median, arising just in front, and a little outside of, anterior eyes.

Tentacular cirri and dorsal cirri of first and second segments, too badly injured to admit of description. Dorsal cirri quite long, reaching to end of the capillary setæ.

Feet same as in *P. tardigrada;* first six with a fan of short, dark colored setæ; capillary setæ from the 7th to the 21st segments inclusive, followed by a few segments without them.

Color : dark blue both above and below, with perfectly black transverse bands, not crossing the ventral surface. Base of feet dark brown. Body filled with large eggs.

Found with *P. tardigrada.* Dredged : 4 to 12 fathoms; on shells, etc.

Fam. NEREIDÆ.

NEREIS (*L.*) *Cuv.*

PL. V, FIGS. 56-61.   PL. VI, FIGS. 65-69.

NEREIS IRRITABILIS *n. sp.*

Body of nearly uniform size throughout the anterior three-fourths; posterior fourth tapering slightly. Number of segments over 200.

Head (f. 56) convex, broadest between the anterior eyes, where the breadth nearly equals the length, decreasing slightly behind, rapidly in front.

Eyes, circular, small, lateral; front pair placed at posterior third; posterior pair near the back of the head, half the diameter of the front pair.

Antennæ, arising close to each other, delicate; length a little more than half the length of the head.

Palpi, about as long as the head, diverging rapidly, extending a little beyond the antennæ; basal portion stout; terminal article slender, elongate.

First segment, as long as the following segments, arching forward, so as to encroach slightly on the head.

Tentacular cirri, 4th reaching to 5th segment; 2d one-third to one-half as long as 4th; 1st and 3d, a little shorter than 2d, equal.

Feet (figs. 58–62), on the anterior part of the body, compact. Superior lingula stout at base, conical, reaching a trifle beyond the upper ramus. The upper lip of the superior ramus is a mere papilla on the anterior basal portion of the lower lip. The lower lip has the same form as the upper lingula, but is a little shorter, and not quite so large at base. The lower ramus is a little shorter than the upper, is divided by a shallow incision into anterior and posterior lips, of which the anterior is slightly the longer, and is further divided at the end into two rounded lobes, upper and lower. Lower lingula, similar to upper, but shorter; after the first few segments projecting a little beyond the lower ramus.

Dorsal cirri, delicate, conical, not quite reaching the end of the upper lingula.

Ventral cirri, similar to dorsal cirri, but shorter; they arise below (within) the base of the lower lingula from a slight elevation, and reach about to the middle of the lower lingula. Further back the upper lip of the dorsal ramus becomes smaller and finally disappears. The posterior lip of the ventral ramus also disappears, and the terminal division of the anterior lip into lobes becomes less marked. The ventral cirri shorten and recede from the lower lingula, while the feet have their basal portion much elongated.

Setæ, all of one kind on the anterior feet (f. 63). Appendix slender, acute, edge beset with numerous short hairs. Terminal points of stem in the same plane (*Aretes homogomphes* Clpd.). They are arranged in three bundles, one in the dorsal ramus, two in the ventral. Ventral setæ very numerous and crowded. Falcate setæ (f. 64) occur first between the 30th and 40th segments; at first in the lower bundle, few in number; becoming more numerous, they occur also in the upper bundle of the lower ramus, but never entirely replace the first form, and disappear again in the posterior segments.

Proboscis (f. 57) about twice the length of the head.

Paragnathi, i, fail; ii, 6 to 10 in small bunches; iii, 2 or 3 irregular transverse rows; iv, 10 to 12 in bunches. All black, small, conical, the inferior (iii), smallest; v, vi, vii, viii, fail.

Jaws, usually light horn color, with dark brown or black edges and tips; curved, twisted externally; denticles small, number variable. In the dorsal ramus there is always one black acicula; in the ventral usually but one, sometimes two or three.

Every specimen found had lost the anal segment.

Dug at low water; mud and sandy mud.

Female. In the adult female the first 33 segments correspond to those described above. The only change consists in a slight enlargement of the eyes. Changes beginning with the 34th segment, and completed at the 40th are as follows (f. 65). At the base of the dorsal cirrus a small membranous plate appears. The lingulæ, and the lower lip of the dorsal ramus, become flattened lanceolate. In the lower ramus the two lobes of the anterior lip are enlarged and flattened, and an additional circular membranous plate appears. At the base of the ventral cirrus a small membranous plate is developed above, and a much larger one below.

At about the 100th segment, a change to the normal form begins. The dorsal cirrus, upper lingula, and dorsal ramus are first affected, next the lower ramus, and last of all the ventral cirrus. This change is nearly as well marked and sudden as in the anterior part.

Setæ: in the changed feet, the ordinary setæ are nearly replaced by cultrate setæ (f. 66), a few, however, remaining in the lower ramus.

Anal segment elongate, striate above.

Anal cirri filiform, in length equal to the last eight segments.

Number of segments, 170 to 220.

Length, 60 to 80^{mm}; greatest breadth, including feet, 5^{mm}.

Color: anterior segments pearl-gray; middle segments blue; base of feet green; posterior third of body gleaming, brassy.

Taken while swimming on the surface from Aug. 1 to Aug. 12. During this time 15 females were taken and

115 males; they were simply dipped up with a scoop-net, indiscriminately, no attempt being made to collect one form rather than the other.

Immature Female. Some specimens dug at low water, July 21, had their bodies well filled with eggs, but were otherwise unchanged, save that the bases of the feet were swollen and elongated.

Color: above pearl-gray, beneath, darker, in some cases light purple. Feet white.

Adult Male. Taken with female. Specimens very numerous, as noted above. The 5th. 6th and 7th dorsal cirri are longer and larger than the others. This is particularly the case with the 7th (f. 68). The 8th cirrus is normal (f. 69).

Change in structure of feet begins on the 34th segment. The development of the membranous plates is better marked in the male than in the female. Those at the bases of the cirri are much larger, and the circular plate on the lower ramus is very large (f. 67).

Color: anterior segments pearl-gray. After the 33d segment, body and base of feet above and below, red; tips of feet white. On each side of the body, both above and below, a narrow white line. Extreme posterior segments sometimes green, sometimes brassy, gleaming; anal segment with a crown of close set papillæ.

NEREIS DUMERILLII *Aud. and M. Ed.*

AUD. and M. ED. Ann. des Sci. Nat., vol. xxix, 1833, p. 218, pl. xiii, figs. 9–12.

(For full description, synonomy, and figures see Ehlers, Borstenwürmer, p. 535; also Claparède, Appendix Ann. Chét. du G. de N., p. 44.)

Three specimens were found, all more or less injured, but certainly belonging to this species. Dredged.

NEREIS VIRENS Sars.

Nereis virens SARS. Beskrivelser og iakttagelser, etc., p. 58, pl. x, f. 27.
" " EHLERS. Borstenwürmer, p. 559, pl. xxii, figs. 29–32.
" " VERRILL. Invert. An. Vin. Sound, p. 590, pl. xi, figs. 47–50.
Nereis grandis STIMP. Invert. of Grand Manan, p. 31, f. 24.
Nereis yankiana QUAT. Hist. des Annelés, vol. i, p. 533, pl. xvii, figs. 7–8.
Alitta virens EXBG. Annulata Nova, p. 112.
" " MGRN. Nord. Hafs-Ann., p. 183.—Annulata Polychæta, p. 172, pl. iv, f. 19.

One specimen only was collected.

NEREIS LIMBATA *Ehlers.*

PL. VI, FIGS. 70–75.

EHLERS. Borstenwürmer, p. 567.
VERRILL. Invert. An. Vin. Sound, etc., p. 590, pl. xi, f. 51; also p. 318.

My specimens, in most particulars, agree perfectly with Ehlers' description, but there are some not unimportant differences. Ehlers states that the dorsal cirri do not extend beyond the lingulæ. This is the case with my specimens until the lingula is enlarged and flattened. Again Ehlers states that the dorsal cirrus, borne by the lingula, is never terminal. I find that on a few of the posterior segments it is terminal (figs. 70–71). I have received from Prof. Verrill specimens referred by him to Ehlers' species which agree perfectly with mine. Verrill (l. c. p. 318) says of the females that "the middle region does not become different from the anterior, as in the male." Up to the time of the reception of the above specimens of adult males, it was my belief that the sexual forms were alike, and that no great change occurred in either. In fact, the changes in the adult male are as well marked as in any species observed by me. Certain changes also occur in the females. The eyes are enlarged so as nearly to touch each other. On the 18th setigerous segment small membranous plates appear at the base of the ventral cirrus, and a similar somewhat larger plate on the lower ramus (f. 73). On the segments thus changed the cultrate (sexual) setæ nearly replace all other forms. In males, with feet changed to the same extent, there were no cultrate setæ. In both male and female, the anal segment is crenulated, and shorter than in the asexual form (f. 71). In both, the feet were much swollen at base. It would not be safe to assume that the changes in the female stop at the point indicated above; for although they seemed to be adult, yet so did the males, and yet Prof. Verrill has shown that the changes in the adult male are much greater than appeared probable from the condition of my specimens. The change in the anal segment occurs first. Many specimens, in which the sexual elements were perfectly distinct, were unmodified, save in the anal segment.

Low water, mud and sandy mud: often in very soft mud, brackish water, being the only annelid that I found living under such conditions. This last station was very near high water mark.

Fam. EUNICIDÆ.

DIOPATRA *Aud. and M. Ed.*

DIOPATRA CUPREA *Clpd.*

Nereis cuprea Bosc. Hist. Nat. des Vers, vol. i, p. 143.
Eunice cuprea QUATR. Hist. Nat. des Annelés, vol. i, p. 331.
Diopatra cuprea CLPD. Annel. Chétopodes du G. de N., p. 432.
" " VERR. Invert. An. Vin. Sound, p. 593, pl. xiii, figs. 67, 68.

I have never seen Bosc's work. The synonomy given above rests on the authority of Claparède.

Common, especially near high water mark.

MARPHYSA *Quat.*

MARPHYSA SANGUINEA *Quat.*

PL. VI, FIGS. 76-80. PL. VII, FIGS. 81-83.

Nereis sanguinea MONTAGU. Linn. Trans., xi, p. 20, pl. iii, f. 1.

Leodice opalina SAV. Syst. des Ann., p. 51.
Nereidonta sanguinea BLV. Dict. Sci. Nat.
Eunice sanguinea AUD. & ED. Ann. Littoral de la France, p. 147.
" " GRUBE. Fam. der Annel., pp. 44, 123.
" " " Die Insel Lussin.
" " " St. Malo and Roscoff.
" " LEIDY. Mar. Invert. Fauna R. I. and N. J., p. 15.
" " JOHNSTON. Cat. of Worms, p. 134.
Marphysa sanguinea QUATR. Hist. Nat. des Ann. vol. i, p. 332, pl. x, f. 1.
" " EHLERS. Borstenwürmer, p. 360, pl. xvi, figs. 8–11.
" " BAIRD. Linn. Proc. Zoölogy, vol. x, p. 352.
" " MAR. & BOBR. Ann. des Sci. Nat., vol. ii, p. 12. 1875.
" *Leidii* QUATR. Hist. Nat. des Ann., vol. i, p. 337.
" *Leidyi* VERR. Invert. An. Vin. Sound, pp. 319, 593, pl. xii, f. 64.

Prof. Leidy first reported this species from our coast. I have no doubt as to the accuracy of his identification. Quatrefages changed the 16 of Leidy's description into 60, in translation, and gave to this new (hypothetical) form a new name, *M. Leidii*. This mistake seems to have passed unnoticed. In regard to Leidy's statement that the branchiæ begin on the 16th segment, it is probable that he examined only small specimens in coming to this conclusion. It will be seen below that the branchiæ may begin on any segment from the 10th to the 23d.

Young Stages: This species abounded in individuals in all stages of growth from 2.5^{mm} to over 20 centimetres in length. The gradation was perfect, and the results seem interesting.

(a.) Youngest form taken: two specimens, 18 and 22 setigerous segments. No antennæ; no branchiae; no indication of division of the head into lobes (palpi) above or below.

Eyes five, situated as in f. 81. These specimens agree with this figure in every particular, except the slight emargination of the head, and in the presence of an antenna.

The feet are fleshy rounded lobes; the dorsal cirri blunt,

finger shaped, extending a little beyond the feet. Ventral cirri swollen at base, outer third suddenly attenuated.

Setæ, in two bundles. Upper bundle made up of two or three compound setæ, appendix short, bidentate (f. 76), and one simple seta (f. 77), long, acute, with a narrow border; one straight black acicula (f. 79). Lower bundle, two or three of the compound bidentate setæ; one compound seta (f. 78), with a long, narrow, acute appendix; one bidentate acicula (f. 80).

(*b.*) In the next stage there is a slight depression of the anterior margin of the head, one antenna, as in f. 81; the bidentate setæ have disappeared from two or three of the anterior segments, the other forms (figs. 77, 78) are more numerous. In other respects same as (*a*). Length, 2.5–4^{mm}.

(*c.*) As a next step, the anterior emargination becomes a little deeper, and a shallow median depression extending back a short distance, above and below (f. 81), indicates more clearly than before the future division of the head into palpi. Simple finger-shaped branchiæ appear on the 10th setigerous segment. On a specimen with 36 setigerous segments there are 13 pairs of branchiæ; of these the first two, and the last three, are shorter than the others.

Length, 4^{mm}.

(*d.*) In the next form, there are three antennæ. In one specimen the lateral antennæ are mere buds (f. 82). Eyes four to two. The median eye disappears first, next one or both of the anterior eye specks, so that on specimens otherwise the same there may be two, three, or four eyes. The division of the head into palpi becomes better marked. The shape of the feet becomes somewhat changed; the bidentate setæ gradually disappear from the anterior segments, until on the largest of the series they are found only on the posterior third; a second straight acicula is added to the upper bundle of setæ, and a second bidentate acicula to the lower bundle. The branchiæ are simple, and begin from the 10th to the 13th setigerous segment. Length, 9^{mm}. Number of segments, 75.

(e.) We have next five antennæ, the external in some cases being minute papillæ, in others having attained their normal (relative) length. On small specimens the division of the head into palpi is not complete. Branchiæ begin on any segment, from the 10th to the 23d, according to the size of the specimen. Eyes, always two.

On specimens from 10 to 20^{mm} in length, having more than 10 and less than 17 non-branchiated anterior segments, the branchiæ are usually bifurcate.

The jaws of the youngest forms correspond perfectly with those of adults, save that the support (Träger) is relatively longer, and with a narrower posterior margin. In this, as in all other respects, a complete gradation can be established.— My figures of jaws all show four teeth on the large denticulated jaw pieces (Zähne, Ehlers), as described by Quatrefages, instead of three, given by Ehlers.

The comb-like setæ of the adult are not found on the youngest forms. I have never seen them on specimens with less than three antennæ, nor on those without branchiæ.

The forms under (c) will be seen to correspond very closely to *Nematoneries oculata* Ehlers (*Borstenwürmer*, p. 374, pl. 16, figs. 19-22), while (d) is very close to AMPHIRO *Knbg.* (*Annulata Nova Öfvers af K. Vet-Akad Förh.*, 1864 p. 565; also *Fregatten Eugenien's Resa*, pl. xvii, f. 26). Kinberg says in his generic diagnosis: "Branchiæ pectiniformes vel subpectiniformes," and for the most part (d) has simple branchiæ, but on one specimen the branchiæ were found divided, the second division being represented by a small but perfectly distinct papilla. Possibly, then,

AMPHIRO may be a young form of MARPHYSA or of some related genus.

DRILONEREIS (*Clpd.*). Char. emend.

Feet uniramous, setæ simple, no ventral or dorsal cirri; lower jaw pieces present, absent, or rudimentary. Upper jaw, on each side, four pieces similar to those on the opposite side, in addition to the support.

Claparède (l. c.) gives as a characteristic of his genus DRILONEREIS: "Labrum nullum." The following species seems to belong to this genus, but it may have two pieces in its lower jaw, or one with the rudiment of the other; or one alone, or finally none at all.

DRILONERIS LONGA n. sp.
PL. VII, FIGS. 84-88.

Head, in the living animal, variable in dimensions; conical, round above, flattened below, bluntly rounded in front.

First two segments equal in length; no appendages.

Feet, on first 30–40 segments, consist of a low rounded elevation, from the summit of which project a few simple, bordered setæ, and one stout acicula. Next appears a minute papilla projecting from the lower back part of the foot; it gradually becomes longer, and extends beyond the foot (f. 84). Meantime the foot itself becomes longer and cylindrical. The posterior lip, which, when it first appears, is about at the outer third of the foot, moves outward, and becomes terminal. Next a papilla appears on the anterior margin of the foot, which gradually elongates and becomes in every respect similar to the posterior lip. These two lips diverge from each other, and between them is the rounded end of the foot from which the setæ project (f. 86). These changes take place very gradually, the feet represented in f. 86, being found only on the posterior third of the body.

Jaws (f. 88). The elongated pieces of the support are not attached in front to a polygonal piece as in *D. filum* Clpd., but coalesce. The posterior pair of maxillary pieces are denticulated at base, but the number of teeth varies from one to five. The part of this piece behind the teeth is very variable in width; it may be nearly double the width shown in figure, and shorter. The second pair of jaw pieces have a peculiar form. Outer margin nearly straight, except in front, where it suddenly curves outward. The first tooth is long, the second quite short, the others of varying length as shown in the figure. The paragnathi or anterior small pieces, are of the same form, the anterior pair a little smaller than the posterior; each consists of a single sharp curved tooth, with two strong curved supporting branches at base.

The lower jaw pieces are extremely variable, both as to form and number; usually their length is about double their greatest width, but their greatest width may be in any part; the opposite pieces are never alike, often one is very minute; sometimes there is but one, sometimes none.

Body: general appearance same as LUMBRICONEREIS. Often a few of the posterior segments have no appendages. Anal cirri four, short. Length variable. One not particularly long specimen had 600 segments, with a length of $23^{cent.}$; diameter 0, 6^{mm}.

Abundant at low water; mud and sandy mud.

LUMBRICONEREIS (*Bl.*) *Ehlers.*

LUMBRICONEREIS TENUIS *Verrill.*

Invert. Animals of Vineyard Sound, etc., p. 591, and p. 342.

Sea Beach. Hog Island.

ARABELLA (*Grube*) *Ehlers.*

Ehlers. Borstenwürmer, p. 398.

Arabella opalina *Verrill.*

Lumbriconereis splendida Leidy. Mar. Inv. Fauna R. I. and N. J., p. 10.
" *opalina* Verrill. Invert. An. of Vin. Sound, etc., p. 594, pl. xiii, figs. 69, 70; also p. 342.

Very common at low water; also dredged frequently.

I have received from Prof. Verrill advance sheets of a check list of Marine Invertebrata of New England, in which the generic position of this species is rectified as above.

STAUROCEPHALUS (*Grube*) *Ehlers.*

Staurocephalus pallidus *Verrill.*

Invert. Animals of Vineyard Sound, etc., p. 595 and p. 348.

Through the kindness of Prof. Verrill, I have been able to compare my specimens with those found by him in the Sound near New Haven. This species is rare in Virginia. I found only five during the entire season. They differ from the Sound specimens only in being narrower; specimens nearly as long as the type forms not being more than 1^{mm} in diameter. The length (50^{mm}) given by Verrill (l. c. p. 596) is not correct. In alcohol the longest of the two original specimens, has a length of 15^{mm}. The antennæ are longer than the head; composed, when perfect, of 12 articles. The palpi (lateral or anterior antennæ) have a stout cylindrical basal part, forming about three-fourths of their entire length, and a terminal fusiform article, separated from the basal part by a deep constriction; they are shorter than the antennæ, but longer than the head. The head is somewhat elongated, more so than in the following species. Eyes small, orange-yellow, the front pair largest. In living specimens the constriction between the basal and terminal parts of the dorsal cirri is very indistinct; in alcoholic specimens it is sufficiently well marked.

The ventral cirri arise at the outer third of foot; at first not extending beyond the foot, but further back, becoming longer, they reach a little beyond the foot.

The upper anal cirri are filiform, composed of 6–8 slender elongated articles. Their length is about that of the last six segments.

The lower cirri are cylindrical, and of the length of the basal article of upper cirri.

The color of my specimens was a uniform yellowish-white. Feet white.

Length, 14^{mm}; diameter, 1^{mm}.

Found on shells, etc. Dredged, 4–12 fathoms.

STAUROCEPHALUS SOCIABILIS n. sp.

Pl. VII, figs. 89, 90, 91.

Head short, the width less than that of the first segment, convex above and below, bluntly rounded in front.

Palpi large, stout, concave below, convex above, margins crenulated; origin visible from below. Externally they taper slightly, and are longer than the head.

Antennæ, longer than palpi, diameter at base about half that of palpi, subulate, composed of 6–10 articles which are quite variable in form. On the inner half the line of division between the articles is indistinct; outer half, distinct. The terminal article is elongate, fusiform.

Eyes circular, dark red, anterior pair very large, situated between the bases of the palpi and antennæ; posterior pair small, just behind and within the bases of the antennæ.

Feet cylindrical, divided at end into upper and lower lips; upper lip further divided into anterior and posterior lobes, all three terminations somewhat flattened, triaugular; lower lip slightly largest. In the posterior half of the body the length of the feet equals the width of the segments; not quite so long in front.

Dorsal cirri, longer than feet: none on first setigerous segment; first shorter than second. Base cylindrical; appendix conical or fusiform, tapers rapidly to a pointed extremity, forms about one-third the entire length of cirrus. Constriction between the base and the terminal article very deep. Blood vessels and seta very apparent in the cirrus.

Ventral cirri, arising at middle line of the foot, do not reach quite to the end of the foot.

Anal cirri: upper made up of four articles, subulate; lower, one article, similar in all respects to basal article of upper cirri.

Body, widest at anterior third, diminishing slightly in front, and more rapidly behind. Convex above, flat below; segmentation well marked.

Setæ: the terminal article of the compound setæ is much shorter and stouter than in *S. pallidus* Verr.; fig. 90 represents a form occasionally seen in the anterior feet. Simple setæ of the upper bundle long, delicate, with one edge minutely denticulated.

Color, yellowish-white; between the segments a narrow red line; a similar line crosses each segment, sometimes falling a little short of the base of the feet. Just outside of each posterior eye is a crescentic red line bounding a clear white spot; while just back of each posterior eye is an irregularly curved red line, also limiting a clear white spot. Anal segment with a quadrangular clear white spot.

Length of a large specimen, 20^{mm}; diameter, 2^{mm}.

Numerous, of all lengths from 4 to 20^{mm}. On shells, etc. Dredged.

Fam. GLYCERIDÆ.

RHYNCHOBOLUS *Clpd.*

RHYNCHOBOLUS AMERICANUS *Verrill.*

Glycera Americana LEIDY. Mar. Invert. Fauna R. I. & N. J., p. 15, pl. xi, figs. 49, 50.
"   "   EHLERS. Borstenwürmer, p. 668, pl. xxiii, figs. 43–46.
"   "   GRUBE. Jahres-Bericht der Schles. Gesell. für vaterlän. Cultur, p. 64. 1869.
Rhynchobolus Americanus VERR. Invert. An. Vin. Sound, p. 596 pl, x. figs. 45, 46.

Common. Low water and dredgings.

RHYNCHOBOLUS DIBRANCHIATUS *Verrill.*

Glycera dibranchiata EHLERS. Op. cit., p. 670, pl. xxiv, figs. 1, 10–28.
"   "   GRUBE. Op. cit., p. 64. 1869.
Rhynchobolus dibranchiatus VERRILL. Op. cit., p. 596, pl. x, figs. 43, 44.

Fam. CHLORÆMIDÆ.

TROPHONIA (*M. Ed.*) *Clpd.*

TROPHONIA ARENOSA *n. sp.*

PL. VII, FIGS. 92–97.

Body elongated, widest at about the 22d segment, tapering gradually in both directions; in front nearly quadrangular, the sides being straight, and the dorsal and ventral surface but slightly convex; after the first 10–12 segments the body is rounded.

Branchiæ very numerous, filiform, red at base, green externally; the inferior shorter than the superior.

Setæ of the first five segments turned forward; elongated, those of the first three reaching far beyond the branchiæ. Tentacles not quite as long as the branchiæ, canaliculated, margins scolloped; color reddish-brown with a green centre. Body above and below with numerous cylindrical papillæ, of which, in adult specimens there are five longitudinal series — one median, with two on each side equally distant from each other. The same number and arrangement holds for the ventral surface. On the anterior segments these papillæ are situated near the anterior margin, and besides those in regular series are numerous smaller ones which project forward from the front of the segments. On young specimens there may be but three series of papillæ.

The pedal rami are far removed from each other. They are surrounded by papillæ similar in all respects to those on the body, in number from 4 to 8. Back of each fascicle of setæ is an elongated cirrus (or papilla) slightly swollen at base; this is somewhat larger and longer in the dorsal than in the ventral rami.

Setæ. The dorsal setæ are all similar to f. 95, save that those of the first five segments and particularly of the first three are very long. The ventral setæ of the first three segments are like the dorsal. On the fourth segment minute terminal hooks appear with a straight tooth below. They shorten backward and the hooks enlarge as in figs. 96 and 97. Dorsal setæ amber-yellow; ventral setæ, at base dark reddish-brown, becoming lighter externally.

Body covered with fine sand, closely adherent.

Fam. CHÆTOPTERIDÆ.

SPIOCHÆTOPTERUS (Sars). Char. emend.

Sars, in his diagnosis of this genus, limits the segments of the middle region of the body to two. The following species agrees so closely with SPIOCHÆTOPTERUS in most respects, that it seems undesirable to form a new genus for it; the middle region, however, is composed of 20–23 segments. Sars also makes the absence of eyes a generic character; but the presence or absence of eyes seems, at least among annelids, never to have more than a specific value.

Spiochætopterus oculatus n. sp.

Pl. VIII, figs. 98-102.

Anterior region composed of the buccal segment and nine setigerous segments. Middle region, 20-23 segments. Posterior region, segments numerous, number variable. Head (f. 98), sunk in buccal segment and prolonged backward between the bases of the tentacles. The anterior part is trapezoidal, a little wider behind than in front; the posterior narrow part has slightly concave lateral and posterior margins, giving rounded, projecting, posterior angles. The greatest width of the head is a trifle less than its length; it is slightly convex. In front it is truncated, or bent directly downward, its front face being somewhat triangular, bluntly rounded at the apex, below.

Buccal segment, forms the anterior part of the body, projecting a little in front of the head; truncate in front, anterior margin thick, rounded; lower and lateral surfaces convex; anterior truncated face concave.

Tentacles, have the usual origin and structure, length 20-25^{mm}.

Eyes two, lateral, just in front of the bases of the tentacles, elliptical, directed obliquely toward the middle line of the head, black.

Anterior region. The feet are flattened, fleshy lobes, thick at base, running out to a thin edge; they are dorsal, transverse, directed upward and a little outward. The setæ arise from the front face of the feet, are long internally, growing progressively shorter externally. The outer setæ do not project beyond the foot. The setæ also change their form from within outward, as shown in f. 101, a, b, c, d. The peculiar broad seta of the 4th setig. segment is slightly convex, has regular transverse markings, widens near the end. The expanded portion is thickened along an irregular line which divides it into two nearly equal parts; its outer margin obliquely truncated, thin, minutely denticulated. The remaining setæ of this segment are of the same form as those found on the other segments.

Middle region. Dorsal ramus, composed of two ventral plates, of which the inner is bilobed, and arises just outside the middle line of the dorsum (f. 99). This plate is connected at base with a similar vertical lateral plate, which, however, is not lobed, and does not contain setæ. The lips of the ventral ramus are anterior and posterior; both are connected with the latero-dorsal plate by a basal membranous prolongation. The posterior lip is a low, arched plate, running downward nearly to the middle line of the ventral surface. The anterior lip is much shorter than the posterior, arising just above and in front of the latter, also a little in front of the latero-dorsal plate. Both lips of the ventral ramus have a yellow color derived from the uncini concealed within them.

Posterior Region. The dorsal rami are vertical (f. 100), swollen at base, otherwise cylindrical, terminating in a button. They are close to the middle line of the body, and contain from four to six very delicate setæ, which project slightly. The ventral rami are the same as in the middle region, only the lips are much reduced in size, and very close to each other.

The anal segment is without appendages, conical, obliquely truncated from above downward, divided below into lobes which can not be seen from above.

Dorsal sulcus, bordered on each side by a delicate raised membrane. These may be revolute, leaving the sulcus open; or their edges may be brought together, completely closing it. Sulcus densely ciliated.

The body tapers uniformly, the diameter of the anal segment being about one-third that of the anterior part.

Anterior region, convex below and laterally; flat or

slightly concave above. Middle and posterior regions rounded.

Color. The anterior region is white above; the first six segments are white below, with numerous brown specks, which often run together, forming blotches, which may occupy more space than the white; remaining segments of this region white. Sometimes the 5th or 6th segment is brownish-purple below, in which case the following one, two, or three segments are flake-white. Middle and posterior regions yellowish-white above; white to light brown below. Between the segments of the last region there is a narrow brown band running down the sides.

Tube. In front white, shining, transparent; further back yellowish-white or brown, sometimes transparent, sometimes opaque. It is annulated with raised bands, which may be black, or white, or of the color of the tube; its diameter 1^{mm}.

Length of animal, $30-50^{mm}$.

Diameter of anterior third, $0.5-0.8^{mm}$.

Found in sand at low water; only in one place, but there abundant.

Fam. SPIONIDÆ.

NERINE *Johnston.*

NERINE HETEROPODA n. sp.

PL. VIII, FIGS. 103-110.

Head, posterior two-thirds rounded (f. 103), swollen; anterior third suddenly acuminate, conical. Separation between head and buccal segment distinct. Buccal segment nearly as long as the next four segments together. The long tentacles arise at the posterior lateral margins of this segment; between their bases are four small black eyes, two on each side, all on the same straight line. Just back of the origin of the tentacles are two fans of setæ, shorter than those on the next segment, but otherwise similar to them. Each fan has a little rounded lobe back of it. These lobes are close together, similar to each other and to the ventral lobes generally, but smaller. The second segment bears a pair of short branchiæ, which increase in length on the following segments, till they nearly touch each other across the back. The anterior branchiæ bear a foliaceous lobe, with a short, free, rounded extremity (f. 104). This lobe is at first divided into two by an incision reaching nearly to the branchia; the lower lobe being double the length of the upper one. Further back the incision disappears, while the free portion of the lobe becomes longer and pointed (f. 105).

Ventral rami. On the anterior feet the ventral ramus is a rounded lobe of the same structure as the branchial lobe. At about the 25th segment, this lobe is nearly divided into two by a deep incision (f. 106); further back this division becomes complete (f. 107) leaving a squarish lobe between the dorsal and ventral rami, not attached to either. The setæ project in front of the lower lobe. On the anterior feet the dorsal setæ fall a little short of the outer margin of the lobe; afterwards they project slightly beyond it. They are numerous, arranged in a fan, delicate, bilimbate, the margins being very narrow. The anterior ventral setæ are like the dorsal (f. 108). At about the 25th segment a few bidentate setæ, with a broad membrane extending about one-third of their length, replace some of the double-margined setæ of the ventral ramus (f. 109). Still further back the bilimbate setæ disappear entirely from the lower ramus, being replaced by narrower setæ with a single margin (f. 110).

Description and figures from alcoholic specimens. Posterior segments not seen. Segments remaining, 40.

Length, 20^{mm}; diameter, 15^{mm}.

POLYDORA Bosc.
POLYDORA HAMATA n. sp.
PL. VIII, FIGS. 111-116. PL. IX, FIGS. 117,118.

Head emarginate in front, outer angles rounded, directed outward and forward.

Eyes four, black, between bases of tentacles; the anterior pair larger and further apart than the posterior.

Branchiæ begin on the seventh segment, 38 pairs, short, flattened, not tapering; bluntly rounded, almost truncate externally.

There is no terminal sucker, but in its place are two somewhat quadrangular lobes (f. 118), each with a short lateral cirrus. Anal opening surrounded by a circle of low papillæ.

Setæ. On the anterior segments are the ordinary capillary setæ, simple acuminate (f. 111); after the change of setæ, these are confined to the dorsal rami. On the fifth segment there are two kinds of setæ; in the upper series from 8 to 10 stout, flattened setæ, slightly curved near the end, not differing much from each other but with no two exactly alike (f. 112); in the lower series from 4 to 6 short capillary setæ, for the most part like f. 111, but sometimes, as in f. 113, widened at outer third.

From 15 to 24 of the posterior segments are furnished with stout hooks, one to each dorsal ramus (f. 116); with these, in addition to the ordinary capillary setæ, is one much longer than the rest (f. 117). The change of ventral setæ takes place on the 7th segment, though a few capillary setæ are found on that segment. The anterior ventral setæ (f. 114) after the change, are somewhat stouter than the posterior (f. 115), all bidentate, covered at the end with a membrane, as shown in the figures.

The anterior part of the body is rounded below and

laterally, flattened above; the posterior third is thin, flattened both above and below.

Color: first third white; middle third brown, semitransparent; last third white; or the anterior third may be light flesh color, posterior two-thirds yellowish; branchiæ red.

Found occupying tortuous galleries in compact bivalve shells. In some respects this form agrees with *P. hoplura* Clpd., but is not identical with it.

POLYDORA CÆCA n. sp.
PLATE IX, FIGS. 119-122.

Head emarginate in front; lateral lobes bluntly rounded, short, no eyes. A rounded carination extends from the head to the anterior margin of the 4th segment.

Tentacles, mere threads; in extension from 10 to 12^{mm} long, white, canaliculated; margins of channel flake-white, interrupted by thirteen transverse dark purple lines, which reach about half-way around the tentacles. Length of markings to spaces included between them about as 1 to 3.

First segment with dorsal ramus only; ramus small and with shorter setæ than those of the following segments.

Second, third and fourth segments with short, blunt, finger-shaped dorsal and ventral cirri.

Branchiæ begin on 7th segment, are of the ordinary form, but very numerous, existing on over 100 segments. The last three are very short, and are followed by sixteen segments without branchiæ.

Terminal sucker funnel-shaped, interrupted below; anal opening with crenulated margin.

Setæ. Very numerous in anterior dorsal rami; some very long, at least one-half longer than the branchiæ. These long setæ for the most part have the form shown in f. 119, but some are narrower. In the upper part of the bundle are shorter setæ (f. 120).

The stout setæ of the 5th segment are bluntly rounded externally, either straight, or slightly curved at tip (f. 121). Three or four short flat capillary setæ curve around the anterior margin of the first stout setæ, and lie directly across them.

Ventral setæ; from the 7th segment stout, bidentate; lower tooth much longer than upper (f. 122), and covered by membrane. They barely project from the surface; are curved within the body.

Color. Anterior part flesh-color passing into white behind.

A single specimen was found living in a tortuous gallery, excavated in a perfectly sound upper valve of *Anomia glabra* Verr. The gallery was lined by a membrane, which formed a tube projecting about 4^{mm} beyond the surface of the shell. The channels in the tentacles were densely covered with cilia. The long tentacles were almost constantly protracted and moved about. The cilia of the branchiæ were very long and numerous.

Length, 19^{mm}; diameter, in front, 1^{mm}; behind, 0.5^{mm}.

This species can readily be distinguished from any previously described from our coast by the purple markings on the tentacles. Though living in a gallery in a hard shell, it lacks the dorsal hooks of *P. hamata* and *P. hoplura* Clpd.

Fam. ARICIIDÆ.

ARICIA (*Sav.*) *Aud. & Ed.*

ARICIA RUBRA *n. sp.*

PL. IX, FIGS. 123-126.

Branchiæ beginning on the 6th setigerous segment.

Head elongate, conical, pointed, as long as the first three segments, having neither eyes, nor antennæ.

Branchiæ: on the anterior segments nearly cylindrical

for the basal three-fourths, becoming suddenly conical externally. They gradually increase in length, and become flattened, triangular, but the outer fourth or fifth remains pointed, conical; on a few of the last segments they again become shorter.

The dorsal rami consist of a delicate finger-shaped dorsal cirrus, shorter than the branchiæ. From a depression at the base of this cirrus a bundle of long simple setæ arises. Further back the dorsal cirri became as long as the branchiæ, and have nearly the same form.

The ventral rami, on the first 25–29 segments, consist of two transverse membranous lips, of which the anterior is very low, hardly perceptible; the posterior a well marked plate, not reaching as far as the setæ, and with a smooth rounded margin. Between these lips are four rows of stout blunt setæ, straight or slightly curved (f. 123). These setæ are all short, first series shortest, increasing progressively in length from first to fourth series. Between the third and fourth series in the upper part of the ramus are a few, usually two, long simple capillary setæ (figs. 124, 125). The ventral rami are short on the first segment, increasing in length from the first to the sixth or seventh segment. At this point they extend from the dorsal to the ventral surface, never encroaching on the ventral surface. The last seven or eight of this series of rami decrease in length progressively, so that the last (25th–29th) is about the same length as the first, and contains about the same number of setæ. The margin of the posterior lip of the last three rami is prolonged into a delicate cirrus.

With the change of setæ the anterior lip disappears; the posterior lip becomes conical, truncated, bearing a cirrus-like prolongation on its lower outer border (f. 126), and its position is changed to the lateral margin of the dorsum.

Between the 40th and 50th feet the dorsal cirri come to be nearly as long as the branchiæ, and of the same form. With the transfer of the ventral rami to the dorsal surface, the feet and branchiæ become connected by an elevated transverse membranous ridge. The dorsal and ventral setæ are alike. The dorsal cirrus stands back of the bundle of setæ which continues to arise at its base, while the ventral setæ arise from the summit and anterior face of the ventral ramus. All the setæ are very long. Throughout most of its length the diameter of the body is uniform. There is a somewhat rapid diminution in the first four segments; and a gradual diminution in the posterior fourth; the last segment having about one-half the width of the middle segments. The anterior segments are slightly concave above (f. 125), slightly convex below; posterior segments (f. 126) flat, or very slightly convex above, broadly rounded at the sides and below, with a distinct median ventral depression.

Anal cirri, four, delicate, subulate, white; the superior of the length of the last five segments; the inferior, short.

Color, red, shade varying in different individuals; middle third sometimes green.

Segments short, numerous.

Length usually about 70^{mm}, with a diameter of 1^{mm}.

Low water, mud and sandy mud.

ARICIDEA n. gen.

One antenna. Feet biramous. Dorsal rami with cirri; ventral rami with cirri on anterior segments only. Branchiæ on anterior segments only. Setæ all simple, capillary. First segment with setæ, no tentacular cirri.

ARICIDEA FRAGILIS n. sp.

PL. IX, FIGS. 127-132.

Head, posterior half convex, sides rounded; anterior half just in front of the single antenna suddenly depressed and narrowed. In contraction this anterior part is about equal in length to the posterior (f. 127) and is somewhat shovel-shaped; in extension it is double this length and is acute, conical. A very shallow incision at the end can be seen in alcoholic specimens; this was not noticed in living forms.

Eyes two, small, circular, lateral, posterior.

Antenna, arises about the centre of the head, delicate, subulate, in length not quite half the head.

Dorsal ramus with dorsal cirrus but without proper feet or ramus. Cirrus on first segment a mere papilla, increasing in length to the fourth segment, where it is about one-third the length of the branchiæ; the first three are cylindrical; the fourth, and all back of it, on the branchiated segments, have a small lobe, developed near the base, looking outward. On the non-branchiated posterior segments this cirrus loses the lobe, becomes somewhat elongated, very delicate, subulate.

Ventral ramus, on first segment a minute cylindrical papilla, a little longer than the dorsal cirrus; increasing slightly in length to the fourth segment, where it is about one-half as long as the dorsal cirrus. Back of the fourth segment it is flattened (compressed) and becomes a fleshy lobe, rounded externally. It retains this form to about the 30th segment, where it again assumes the form of a cirrus, shortens progressively, and disappears at about the 40th segment, after which segment the ventral rami are entirely without appendages. The ventral cirri are always just back of the ventral setæ.

Branchiæ, begin on fourth segment, arise just within the dorsal cirri, taper slightly to near the end when they became suddenly acuminate; they are usually directed inward and a little upward, their length being a little more than half the width of the body, so that they cross each other. The first two or three are a little shorter than those following them. They exist on 50 to 55 segments, the last two or three having about the same length as the first. After the first few segments they entirely cover the back. The anterior dorsal setæ form a fascicle just in front of the dorsal cirrus, they are very long, (f. 131), regularly acuminate. In each bundle there are two or three in the upper part double the length of the one figured. Back of the branchiated segments, there are only four to eight in each bundle, which are a little shorter and more delicate than the anterior setæ.

Ventral setæ (f. 132, a, b.), arranged in three transverse series, numerous, inner third or half of uniform width, then suddenly acuminate, capillary termination very long and delicate. In the upper part of each ramus are a few setæ double the length of those figured; back of the branchiæ they form two fan-like series.

On the anterior branchiated segments, the dorsal and ventral surfaces of the body are very slightly convex. Their convexity progressively increases (f. 129) backwards; the sides are rounded (f. 128). On the non-branchiated segments, the lateral and ventral surfaces are regularly rounded, but there is a sudden contraction of the dorsum just above the dorsal rami (f. 130). Posterior segments not found.

A fragment 32^{mm} in length had 170 segments, and was 2^{mm} in diameter.

Color: the head is white; first 30–40 segments bright red, passing into white, which further back passes into a green. Anterior dorsal setæ, amber-yellow; anterior ventral, brownish-yellow; all setæ back of branchiæ gleaming white.

Rare; low water, mud.

ANTHOSTOMA *Schmarda.*

ANTHOSTOMA ROBUSTUM *Verrill.*

Invert. Animals of Vineyard Sound, etc., p. 597, pl. xiv, f. 76.

One specimen, much injured, but probably referable to this species.

ANTHOSTOMA FRAGILE *Verrill.*

Invert. Animals of Vineyard Sound, etc., p. 598.

A single injured specimen.

Fam. CIRRATULIDÆ.

CIRRATULUS *Lamarck.*

CIRRATULUS GRANDIS *Verrill.*

Invert. Animals of Vineyard Sound, etc., p. 606, pl. xv, figs. 80, 81.

Common both in sand and mud, low water.

Fam. CAPITELLIDÆ.

ANCISTRIA *Quatrefages.*

ANCISTRIA MINIMA *Quatr.*

Hist. Nat. des Ann., vol. ii, p. 252, pl. xi, figs. 28-34.

Claparède remarks that ANCISTRIA *Quatr.* is founded on a CAPITELLA. I know very little indeed about this group, but am certain that my specimens belong to ANCISTRIA as defined by Quatrefages and have found it impossible to separate them from *A. minima.*

Fam. MALDANIDÆ.

CLYMENELLA *Verrill.*

CLYMENELLA TORQUATA *Verrill.*

Clymenella torquata VERR. Invert. An. Vin. Sound, etc., p. 608, pl. xiv, figs. 71-73.

Clymene torquatus LEIDY. Marine Invert. Fauna R. I. & N. J., p. 14.

Low water, sandy mud; common.

MALDANE (*Gr.*) *Mgrn.*

MALDANE ELONGATA Verrill.

Invert. Animals of Vineyard Sound, etc., p. 609.

Common in mud and sandy mud, near low water mark.

Fam. AMMOCHARIDÆ.

AMMOCHARES *Grube.*

Only one specimen belonging to this genus was found, and that having only a few anterior segments, not sufficient for identification.

Fam. HERMELLIDÆ.

SABELLARIA *Lam.*

SABELLARIA VARIANS *n. sp.*

PL. IX, FIGS. 133-136. PL. X, FIGS. 137-139.

Body stout, diameter decreasing regularly from the first segment. Operculum surrounded at base by a series of conical papillæ. On the dorsal surface the inner papilla on each side is flattened at base, and between it and the operculum are a few golden setæ, curved to follow the outline of the opercular base.

The tentacles have the usual structure, are white, with sometimes a red or purple tinge.

First segment not visible above, represented below by two rounded lobes and two setigerous processes or ventral rami. The lobes are placed one on each side of the mouth; outer margin convex, inner margin concave. In life they are continually in motion, being alternately straightened or moved from the mouth, and curved inward to the mouth. Just outside of these lobes is a conical process, arising from a swollen irregularly shaped base. On the outer surface of the base is a curved depression, from which a series of setæ arises, similar to f. 139, but with the lateral appendages much longer than in that figure. The setæ and the process itself point downward.

The second segment bears branchiæ and laterally an acutely conical process with a flattened triangular base; a bundle of setæ, similar to those of the first segment, arises from the lower margin of this process near its base. There are no other setæ on this segment. Accordingly, the first two segments are uniramous having only the ventral rami and setæ. The pedal ramus of the second segment is very close to the setigerous lobe of the first, being on the antero-lateral margin of its own segment. Without careful examination it would seem that both rami belonged to the first segment.

The dorsal setæ of the third segment (f. 136) are smaller, but do not differ otherwise from those of the fourth and fifth (f. 137). Those following the fifth segment have four sharp teeth increasing progressively in length (f. 138). Branchiæ, about twenty pairs; sixth or seventh largest, growing smaller gradually, the last being quite short. Length of anterior branchiæ a little more than the width of the body.

Color: opercular lobes white with dark brown specks; first three or four segments black laterally sometimes ventrally; body generally white, often with a greenish tinge: branchiæ white or red with green centre: caudal appendage green: feet generally white, with sometimes the outer border of the posterior dorsal rami black. In a well marked color variety, the first four segments were entirely black; body green with brown spots; middle third, ventral surface, dark brownish-red; feet light green at base, passing into dark brown externally.

Length, from 8 to 18^{mm}.

The tubes of this species were numerous on shells, etc., dredged from four to twelve fathoms.

Fam. AMPHICTENIDÆ.

PECTINARIA *Lam.*

(Subgenus LAGIS *Mgrn.*)

PECTINARIA (LAGIS) DUBIA *n. sp.*

PL. X, FIGS. 140-144.

Two specimens were found which seem to belong to Malmgren's genus or subgenus LAGIS. Through some carelessness I failed to make notes on them when they were collected. Their present condition does not admit of specific description, except as regards the setæ. These, in some respects are peculiar; there are eight paleolæ on each side, flattened, slightly curved to near the end, terminating in a stout hook (f. 140). The dorsal setæ are of two forms: one, simple, acuminate (f. 141); the other, also simple, but curved slightly, and with a large, well defined projection or knee at about the outer fourth (f. 142), beyond which the edge is divided into numerous fine, hair-like, projections by oblique incisions. Ventral uncini (f. 143) with seven long, sharp teeth.

Spinulæ of the scapha (setæ of the caudal process) in two series, ten in each; of which seven are of the form a, b, f. 144, and three correspond to c, d, e.

Length of one specimen 6^{mm}; of the other 4^{mm}.

Fam. AMPHARETIDÆ.

MELINNA *Mgrn.*

MELINNA MACULATA *n. sp.*

PLATE X, FIGS. 145-147.

Branchiæ green with red centre, and narrow transverse white bands. Tentacles not much longer than branchiæ, light flesh-color.

Body flesh-color, sometimes tinged with green and with numerous flake-white specks on the anterior dorsal surface, nine or ten posterior segments dark brown to black. Eighteen anterior segments, with capillary setæ; posterior segments, 55. There is a narrow raised ventral band between the bases of the opposite feet on the first 17 setigerous segments. The first three setigerous segments have a narrow white band on which the setæ stand, running down the sides, becoming wider and rounded in front.

The spinulæ are slightly convex, strongly striated transversely; longitudinal striæ interrupted by the transverse; of uniform width to near the end, when they curve suddenly, ending in a sharp point (f. 147).

Capillary setæ bilimbate; one margin very narrow and shorter than the other (f. 146). Uncini with four long sharp teeth, and one short tooth (f. 145).

This species is certainly closely related to *Sabellidis* (*Melinna* Mgrn.) *cristata* Sars, but probably not the same. It will be seen that the uncini agree with those figured by Sars (*Fauna littoralis Norvegiæ*, pl. ii, figs. 6, 7), except that they have an additional tooth. The spinulæ are also very much alike (compare f. 147 with Sars l. c. pl. ii, f. 5).

Malmgren has figured the same parts (*Nord. Hafs-Ann.*, pl. xx, figs. 50 D, 50 C), but his figures differ so much from those given by Sars that it hardly seems probable they were made from the same species.

Fam. TEREBELLIDÆ.

AMPHITRITE (*Müller*) *Mgrn.*

AMPHITRITE ORNATA *Verrill.*

Terebella ornata LEIDY. Marine Invert. Fauna. R. I. and N. J., p. 14, pl. xi, figs. 44, 45.

Amphitrite ornata VERR. Invert. An. Vin. Sound, etc., p. 613, pl. xvi, f. 82.

Very common in mud and sandy mud, at low water.

SCIONOPSIS *Verrill.*

SCIONOPSIS PALMATA *Verr.*

Invert. Animals of Vineyard Sound, etc., p. 614.

The few examples which were found, were of small size.

PISTA *Mgrn.*

PISTA CRISTATA *Mgrn.*

MALMGREN. Nord. Hafs–Ann., p. 382, pl. xxii, f. 59.
VERRILL. Amer. Journ. Sci. Arts, Third Ser., vol. x, p. 40.

A single imperfect specimen was collected, consisting of the anterior half.

LEPRÆA *Mgrn.*

LEPRÆA RUBRA *Verrill.*

Invert. Animals of Vineyard Sound, etc., p. 615.

Common, but all of the specimens were of small size.

POLYCIRRUS (*Grube*) *Mgrn.*

POLYCIRRUS EXIMIUS *Verrill.*

Torquea eximia LEIDY. Marine Invert. Fauna R. I. & N. J., p. 14, pl. xi, figs. 51, 52.
Polycirrus eximius VERR. Invert. An. Vin. Sound, etc., p. 616, pl. xvi, f. 85.

ENOPLOBRANCHUS *Verrill.*

ENOPLOBRANCHUS SANGUINEUS *Verr.*

Chætobranchus sanguineus VERR. Invert. An. Vin. Sound, etc., p. 616.
Enoplobranchus sanguineus VERR. Check List of Marine Invert. of the Atlantic Coast from Cape Cod to the Gulf of St. Lawrence, p. 10 (advance sheets).

LYSILLA *Mgrn.*

LYSILLA ALBA *n. sp.*

PLATE X, FIG. 148.

Frontal membrane divided into five lobes by deep incisions, each lobe being scolloped or deeply folded.

Cirri, long, white with colorless centre; some filiform, others enlarged and flattened in their outer third.

Body, first third swollen, then tapering uniformly but very slightly. Anterior two-thirds white, transparent; internal organs showing through. Alimentary canal surrounded, except for a short distance in front, by a yellow glandular mass which is divided on each side of the middle line into semicircular lobes or plates of varying thickness. First third may be regarded as made up of fourteen or fifteen segments, but the segmentation is very obscure. On most specimens there are peculiar bodies on thirteen segments, beginning with the second, appearing as lateral circular elevations with raised margin, concave, centre occupied by a minute elevated point. I know nothing as to the function of these bodies which are not found on some specimens. No segmentation could be made out along the middle third of fresh specimens. In alcoholic specimens, on the ventral surface, lines of division can be seen, but they are very obscure. The posterior third is light gray, with brown specks, opaque, and terminates in a clear white anal segment.

Segments clearly defined, short, numerous. The anal aperture has a minutely crenulated margin.

The entire body is covered with narrow, raised transverse lines, made up of minute verruciform bodies; these are less apparent on the middle third than elsewhere. In living specimens a narrow, raised, flattened band occupies the middle line of the ventral surface. In alcoholic specimens the ventral surface is much depressed, and an elevated rounded ridge on each side separates it from the lateral surface.

Setæ. After careful examination of many specimens I believed that this species was without setæ. Afterwards I found a single fascicle of six setæ, one of which is represented in f. 148. Malmgren assigns to LYSILLA six setigerous segments. I am unable to say how many segments of this species are setigerous, but have referred it to LYSILLA because in all other respects it agrees with that genus.

Fam. SABELLIDÆ.

SABELLA (*L.*) *Mgrn.*

SABELLA MICROPHTHALMA Verrill.

Invert. Animals of Vineyard Sound, etc., p. 618.

POTAMILLA *Mgrn.*

POTAMILLA TORTUOSA *n. sp.*

PLATE X, FIGS. 149-153.

Branchiæ, six pairs, colorless, transparent, with transverse markings of brown and white, and a green centre, connected at base for a short distance by a membrane. Pinnæ often with same markings as stem. The first pair (dorsal) of branchiæ are without eyes. The 2d, 3d, and 4th may have from one to three convex dark red eyes, which are placed at equal distances from each other, the outer one being about half-way out on the stem; their transverse diameter equals the width of the stem; longitudinal diameter a little longer than the transverse. Fifth and sixth pairs of branchiæ without eyes.

The ventral sulcus is continued on the dorsum, but is hardly perceptible save on the first three segments. Anterior part of body usually composed of eight segments, but may have seven, eight or nine. Each of these segments has its ventral surface divided into two equal parts by a transverse impressed line. Back of these the ventral sulcus divides the ventral surface of each segment into lateral quadrangles, whose length is double their width in front; the length becoming less as the segments shorten, they finally become squares. The tentacles are flattened, triangular at base, becoming subulate further out; length, about one-third that of the superior branchiæ.

Pinnæ long, densely ciliated.

All the setæ of the first segment are capillary, similar to f. 149, but with narrower margin. The manubrium of the anterior uncini is very long (f. 152). The short capillary setæ of the anterior segments (f. 150), evidently agree better with LAONOME than with POTAMILLA. The ordinary form of capillary setæ on the posterior segments is shown in f. 153, but there are a few with narrower margins, and also some with a single margin, similar to f. 149. Body of nearly uniform size, a few of the last segments tapering slightly.

This species lives in tortuous galleries excavated in compact shells, lined with a delicate membrane, which projects from 6 to 10^{mm}. No entire specimen was obtained. They could not be withdrawn without breaking, and it was difficult to follow the galleries, owing to their tortuous course and to the compactness of the shells.

A nearly entire specimen, having 100 segments, was 15^{mm} in length; width not quite 1^{mm}. Color, reddish-brown above, white below.

This species lives in colonies, and scattered among them were frequently found individuals with only three pairs of branchiæ, light red, without eyes, quite short and narrow. As in all other respects they agree perfectly with the species described above, they may be the young of that form.

I have referred this species to POTAMILLA *Mgrn.*, although it does not agree with that genus in all respects. The ventral sulcus is certainly continued on the dorsum. The form of the shorter capillary setæ has been already referred to. Aside from these points it agrees with POTAMILLA.

Fam. SERPULIDÆ.

HYDROIDES *Gunnerus.*

HYDROIDES DIANTHUS *Verrill.*

Serpula dianthus VERR. Invert. An. Vin. Sound, etc., p. 620.
Hydroides dianthus VERR. In Notes on the Nat. Hist. of Fort Macon, N C., Coues and Yarrow (No. 5), in Proc. Acad. Nat. Sci. Phila., 1878 (advance sheets).

GENERA INCERTÆ SEDIS.

CABIRA n. gen.

Sides of head produced into thin plates, which are covered with papillæ. First segment with two pairs of tentacular cirri, without setæ. Dorsal cirri on all segments; no ventral cirri. Ventral setæ, stout hooks beginning on the 6th setigerous segment, one to each ramus.

CABIRA INCERTA n. sp.

PLATE XI, FIGS. 155-157.

Head convex above and below, bluntly rounded in front, lateral margins prolonged into membranous expansions, which project in front of the head, and are densely covered with minute papillæ. Antennæ two, minute, arising from a short swollen base, lateral, situated at the anterior third (f. 154).

Buccal segment, in length about equal to the two following segments; no setæ; two minute tentacular cirri on each side; separated from the following segment by a deep lateral constriction.

Feet. Dorsal ramus with a small dorsal cirrus; similar in all respects to the antennæ and tentacular cirri; beneath this arise from two to four delicate capillary setæ (f. 157, a, b). Ventral ramus a small papilla, from the summit of which issues a strong hooked seta (f. 156). The first five setigerous segments have only the dorsal rami.

Body convex above, flat below. Posterior part not seen.

Length of part examined (40 segments), 12^{mm}.

Color, light gray.

Found living in a fragment of loosely compacted sandstone. Dredged.

PHRONIA *n. gen.*

Head divided into palpi. Body elongate, flattened, composed of numerous segments. First segment with two pairs of tentacular cirri, without setæ. Dorsal cirri of second segment similar to the upper tentacular cirri. Remaining dorsal cirri flattened, thin. Feet uniramous. Setæ all simple, capillary. Anal segment not seen.

PHRONIA TARDIGRADA *n. sp.*

PLATE XI, FIGS. 158-163.

Head (f. 158), width greater than length; lateral margins slightly concave; posterior margin straight to the middle third, then curving suddenly backward, encroaching on the front of the buccal segment; on the lower surface, close to the inner anterior margin of the frontal lobes (palpi), two minute papillæ, which, pointing forward and inward, project a little beyond the palpi; in front the division between the palpi extends about one-third of the way back; above, is continued as a shallow depression; below, as a deeply impressed line, to the posterior margin: palpi broadly rounded in front. Antennæ two, conical, lateral, very small.

Tentacular cirri; superior, conical, elongate, turned forward, reaching a little beyond the head: inferior, conical, minute. Sides of the buccal segment turned forward, embracing the head. The mouth has a crenulated posterior margin. Dorsal cirri of the second segment similar to those of the first, but a little longer and very slightly flattened; after the first they have a cylindrical or slightly tapering basal article; back of the second segment they are suddenly shortened and flattened, becoming leaf-shaped (figs. 159, 160). The substance of these cirri is crowded with small glandular bodies, and those of the anterior segments have a series of minute cylindrical papillæ along their inner back margin. The ventral cirri originate at first near the base of the dorsal cirri (f. 159); further back that part of the basal article carrying the dorsal cirrus becomes longer, but the position of the ventral cirrus remaining unchanged (f. 162), it now falls short of the base of the dorsal cirrus. The middle third of the anterior segments is slightly convex above, otherwise the body in this region is flat both above and below; further back the outline is that shown in f. 164. The posterior setæ (f. 161) are longer and stouter than the anterior but not so numerous.

In the single specimen found there were 320 segments; the posterior portion of the body had been lost.

Length observed, 90^{mm}.

Diameter of head, 0.4^{mm}; breadth of third segment, 1.1^{mm}; of 100th segment, 4^{mm}; of 300th segment, 3.5^{mm}.

All the body measurements include the feet; and the changes in width are almost entirely due to the increase, and subsequent decrease, in the length of the feet; the width of the body changing but very little.

Color everywhere white.

Found at extreme low water in soft mud.

INDEX TO THE ANNELIDA CHÆTOPODA.*

Alitta virens,	235	Laonome,	266
AMMOCHARIDÆ,	259	Leodice opalina,	237
AMMOCHARES,	259	LEPIDAMETRIA,	209
AMPHARETIDÆ,	261	Lepidametria commensalis,	210
AMPHICTENIDÆ,	261	Lepidonote armadillo,	204
AMPHIRO,	239	punctata,	204
Amphitrite ornata,	213, 262	Lepidonotus squamatus,	204
Ancistria minima,	258	Augustus (Var.),	205
Antinoë parasitica,	208	variabilis,	205
Anthostoma fragile,	258	Lepræa rubra,	263
robustum,	258	Lumbriconereis opalina,	242
Aphrodita punctata,	204	splendida,	242
squamata,	204	tenuis,	241
Arabella opalina,	242	Lysilla alba,	263
Aricea rubra,	253	Maldane elongata,	259
ARICIDEA,	255	MALDANIDÆ,	258
fragilis,	255	Marphysa Leidii,	237
ARICIIDÆ,	253	Leidyi,	237
AUTOLYTUS,	224	sanguinea,	236, 237
hesperidum,	225	Melinna cristata,	262
ornatus,	225	maculata,	261
CABIRA,	267	Nematonereis oculata,	239
incerta,	267	NEPHTHYDIDÆ,	213
CAPITELLIDÆ,	258	Nephthys ingens,	213
Chætobranchus sanguineus,	263	picta,	214
CHÆTOPTERIDÆ,	246	NEREIDÆ,	231
CHLORÆMIDÆ,	245	Nereidonta sanguinea,	237
CIRRATULIDÆ,	258	Nerine heteropoda,	249
Cirratulus grandis,	258	Nereis Dumerillii,	234
Clymene torquatus,	258	grandis,	235
Clymenella torquata,	258	irritabilis,	231
Diopatra cuprea,	236	limbata,	235
DRILONEREIS,	240	Nereis sanguinea,	236
longa,	240	virens,	235
Enoplobranchus sanguineus,	263	yankiana,	235
Eumida maculosa,	215	Odontosyllis Dugesiana,	220
Eunice cuprea,	236	fulgurans,	220
sanguinea,	237	PÆDOPHYLAX,	223
EUNICIDÆ,	236	dispar,	223
HALOSYDNA,	210	Pectinaria dubia,	261
HERMELLIDÆ,	259	Pista cristata,	263
HESIONIDÆ,	216	Phyllodoce fragilis,	214
Hydroides dianthus,	266	PHYLLODOCIDÆ,	214
Glycera Americana,	245	PHRONIA,	268
dibranchiata,	245	tardigrada,	268
GLYCERIDÆ,	245	Podarke obscura,	216
Lagis dubia,	261	Polycirrus eximius,	263

* For pages in the Advance Copies subtract 200.

72 Index.

Polydora cæca, ... 252
  hamata, ... 251
Polynoë dasypus, ... 204
  squamata, ... 204
POLYNOIDÆ, ... 204
Potamilla tortuosa, ... 265
PROCERÆA, ... 224, 227
  cœrulea, ... 230
  luxurians, ... 225
  macrophthalma, ... 225
  ornata, ... 225
  picta, ... 225
  tardigrada, ... 227
Rhynchobolus Americanus,... 245
  dibranchiatus, ... 245
Sabella microphthalma,... ... 265
Sabellaria varians, ... 259
SABELLIDÆ, ... 265
Sabellides cristata, ... 262
Scionopsis palmata, ... 262
Serpula dianthus, ... 266

SERPULIDÆ, ... 266
SIGALIONIDÆ, ... 213
SPHÆROSYLLIS, ... 222
  Claparèdii, ... 223
  fortuita, ... 221
  hystrix, ... 223
  pirifera, ... 222
SPIOCHÆTOPTERUS, ... 246
  oculatus, ... 247
SPIONIDÆ, ... 240
Staurocephalus pallidus, ... 242
  sociabilis, ... 243
Sthenelais picta, ... 213
SYLLIDÆ, ... 217
Syllis fragilis, ... 217
  gracilis, ... 217
Terebella ornata, ... 262
TEREBELLIDÆ, ... 262
Torquea eximea, ... 263
Trophonia arenosa, ... 245

EXPLANATION OF PLATE I.

LEPIDONOTUS SQUAMATUS Kbg.
Page 304.

Fig. 1. Head of a small specimen, × 25.
Fig. 2. Foot seen from below, × 25.
Fig. 3. Seta of ventral ramus, × 140.
Fig. 4. Seta of dorsal ramus, × 240.
Fig. 5. Seta of first ventral ramus, × 240.

LEPIDONOTUS VARIABILIS n. sp.
Page 205.

Fig. 6. Head of a small specimen, × 25.
Fig. 7. Elytron from middle of body, × 15.
Fig. 8. Foot from a larger specimen, × 25.
Figs. 9, 10. Setæ of lower ramus, × 140.
Fig. 11. Seta of first segment, lower ramus, × 140.

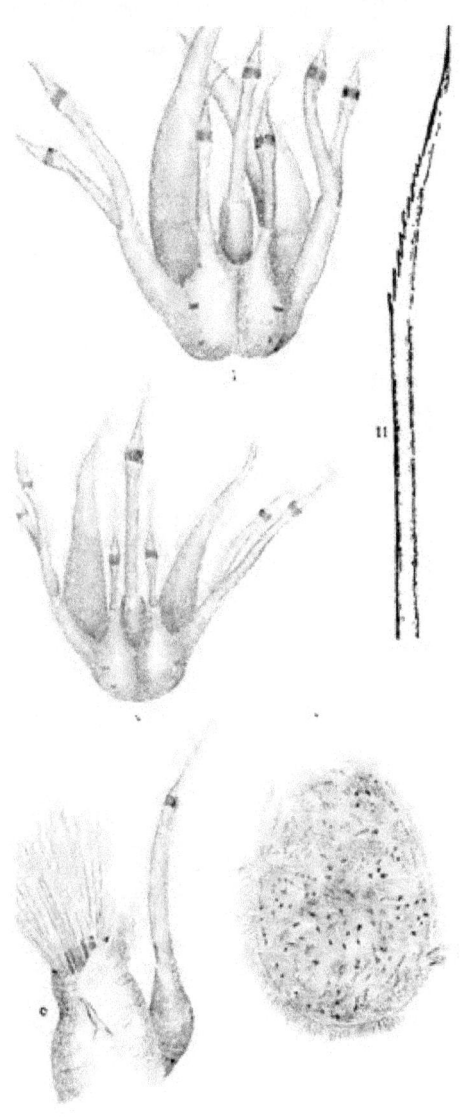

EXPLANATION OF PLATE II.

LEPIDONOTUS VARIABILIS n. sp.
Page 205.

Fig. 12. Seta of dorsal ramus, × 410.
Fig. 13. Seta of ventral ramus, worn, × 140.
Fig. 14. Head of ? sexual form, × 25.

ANTINOË PARASITICA n. sp.
Page 208.

Fig. 15. Head, enlarged.
Figs. 16, 17. Elytra, × 35.
Fig. 18. Seta of dorsal ramus, × 460.
Fig. 19. Seta of ventral ramus, × 410.
Fig. 20. Seta of ventral ramus, first two segments, × 460.
Figs. 21, 22. Hooked setæ of last two segments, × 410.

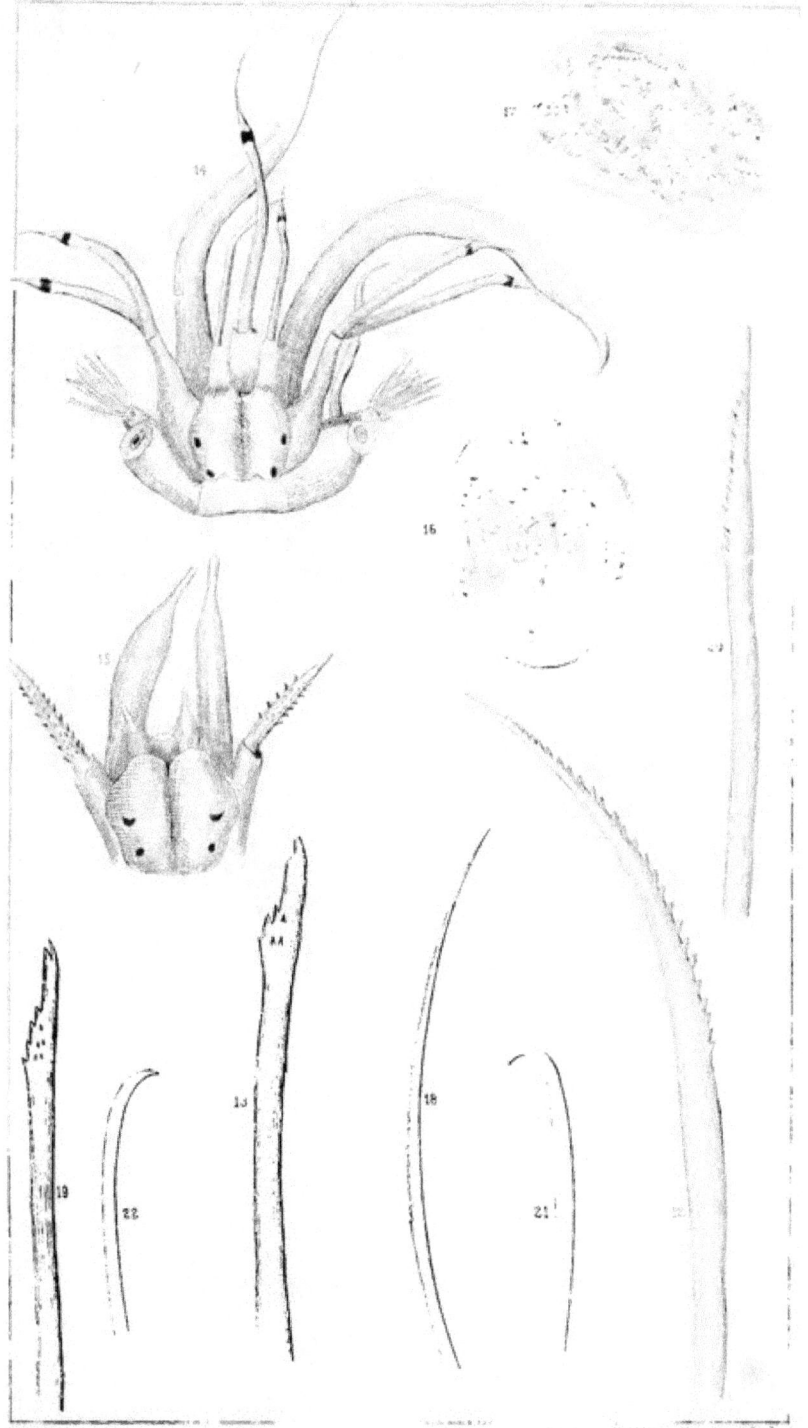

EXPLANATION OF PLATE III.

LEPIDAMETRIA COMMENSALIS n. sp.
Page 210.

Fig. 23. Head, × 25.
Fig. 24. Jaw piece, × 25.
Fig. 25. Emarginate elytron, × 5.
Fig. 26. Oval elytron, × 5.
Fig. 27. Posterior elytron, × 5.
Fig. 28. Foot, × 25.
Fig. 29. Large pointed seta, × 70.
Fig. 30. Bidentate seta of lower ramus, × 140.
Fig. 31. Seta of dorsal ramus, × 240.

PHYLLODICE FRAGILIS n. sp.
Page 214.

Fig. 32. Head and anterior segments, × 35.
Fig. 33. Foot from anterior third, × 30.
Fig. 34. Foot from posterior third, × 30.
Fig 35. Foot from a young specimen, × 30.
Fig. 36. Seta, × 750.
Fig. 37. Foot with ventral cirrus, × 30.

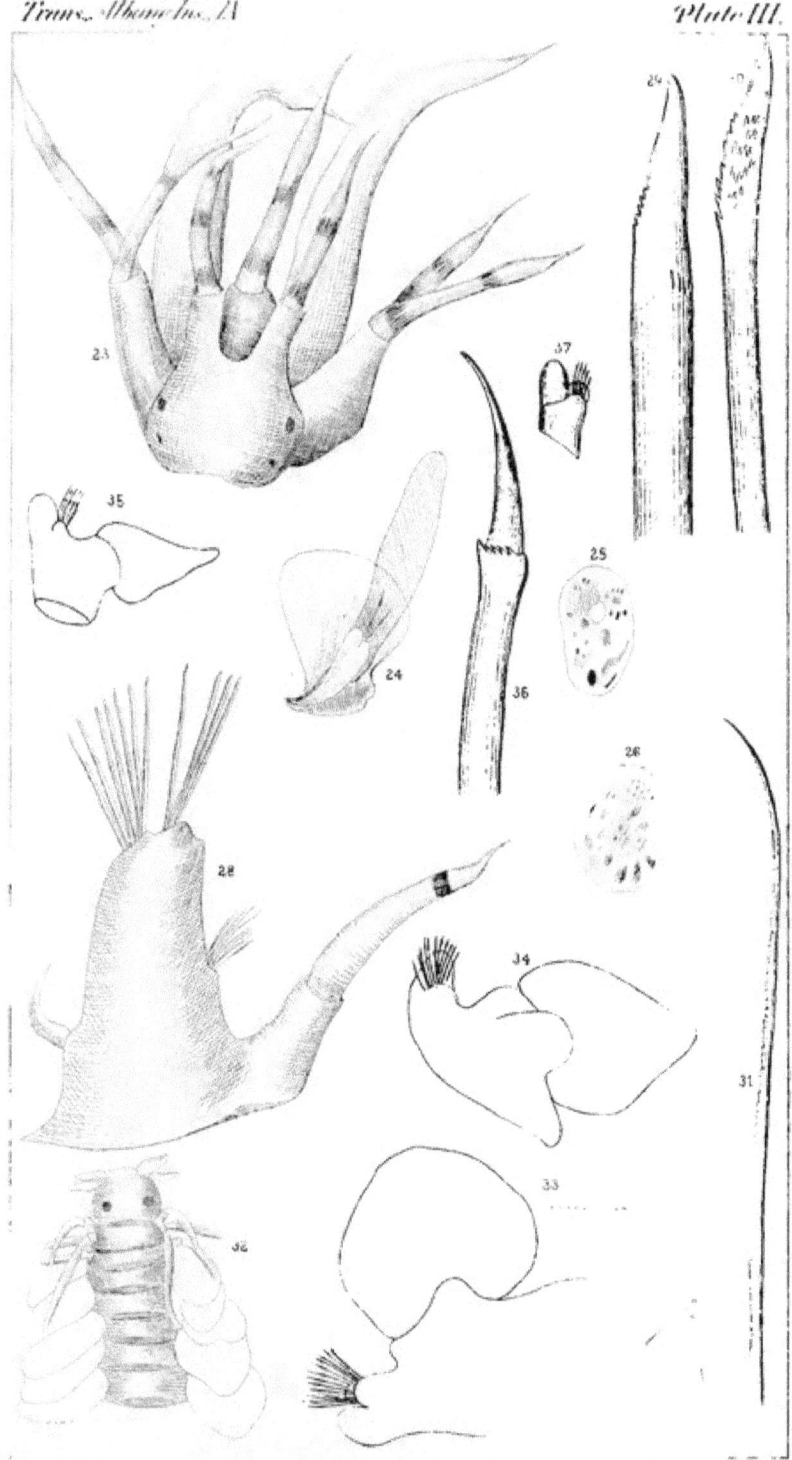

Plate III.

EXPLANATION OF PLATE IV.

EUMIDA MACULOSA n. sp.
Page 215.

Fig. 38. Head, etc., × 70.
Fig. 39. Anterior foot, × 70.
Fig. 40. Foot from middle third, × 70.
Fig. 41. Posterior dorsal cirrus from a large specimen, × 70.

SYLLIS FRAGILIS n. sp.
Page 217.

Fig. 42. Seta, × 750.
Fig. 43. Acicula, × 750.
(Fig. 43 is not good, the terminal button not being shown.)

SPHÆROSYLLIS FORTUITA n. sp.
Page 221.

Fig. 44. Head and anterior segments, × 130.
Fig. 45. Posterior segments, × 130.
Fig. 46. Compound seta with long appendix, × 750.
Fig. 47. Compound seta with short appendix, × 750.
Fig. 48. Simple seta, × 750.

PÆDOPHYLAX DISPAR n. sp.
Page 223.

Fig. 49. Head and anterior segments, × 130.

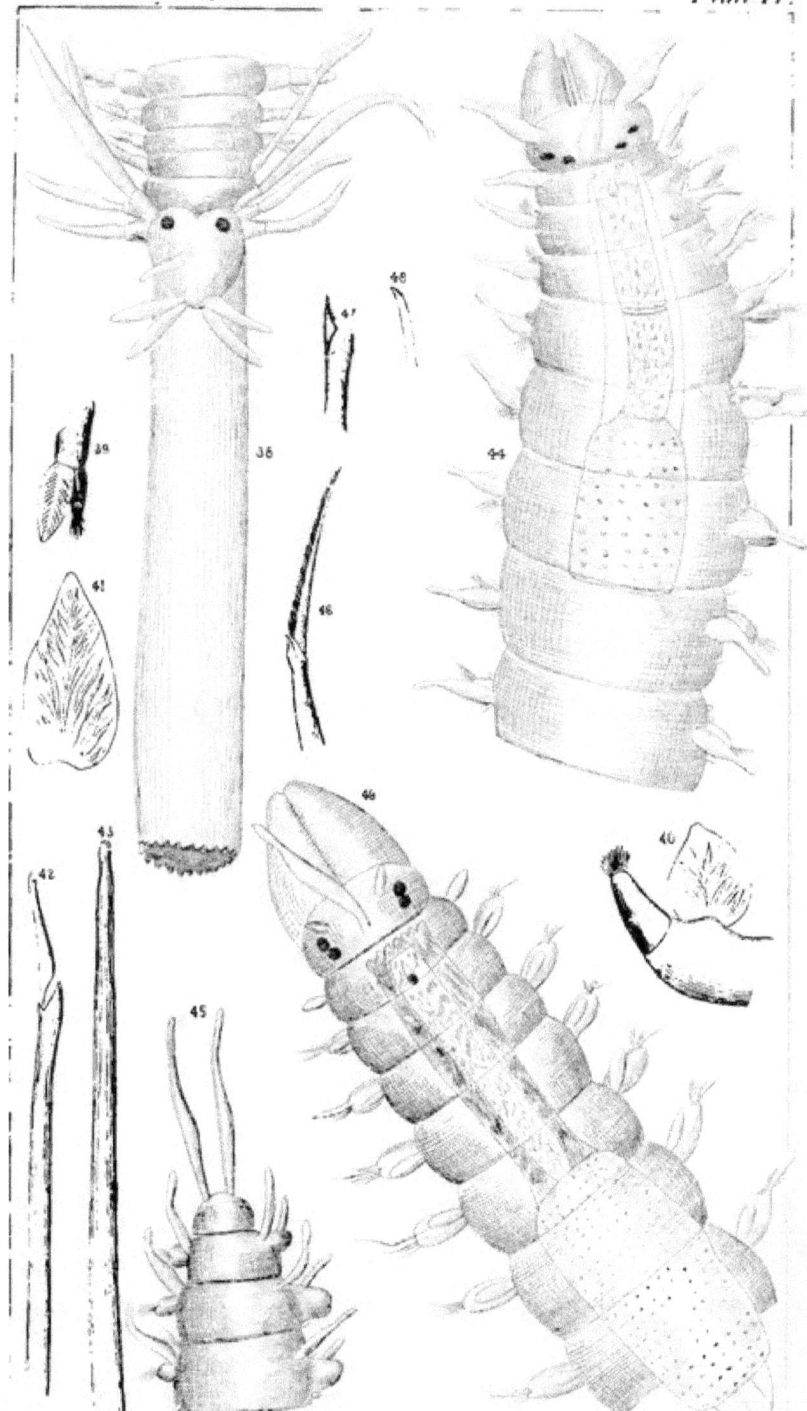

EXPLANATION OF PLATE V.

PÆDOPHYLAX DISPAR, n. sp.
Page 223.

Fig. 50. Posterior segments, × 130.
Fig. 51. Seta with short appendix, × 750.
Fig. 52. Seta with longer capillary appendix, × 750.
Fig. 53. Same as 52, different view, × 750.
Fig. 54. Simple seta, slightly curved, × 750.
Fig. 55. Simple seta, more curved, × 750.

NEREIS IRRITABILIS n. sp.
Page 231.

Fig. 56. Head and extended proboscis, × 20.
Fig. 57. Proboscis, ventral view, × 20.
Fig. 58. Foot from 7th segment, × 20.
Fig. 59. Foot from 30th segment, × 20.
Fig. 60. Foot from 50th segment, × 20.
Fig. 61. Foot from 70th segment, × 20.
Fig. 62. Foot from 160th segment, × 20.
Figs. 63, 64. Setæ, × 450.

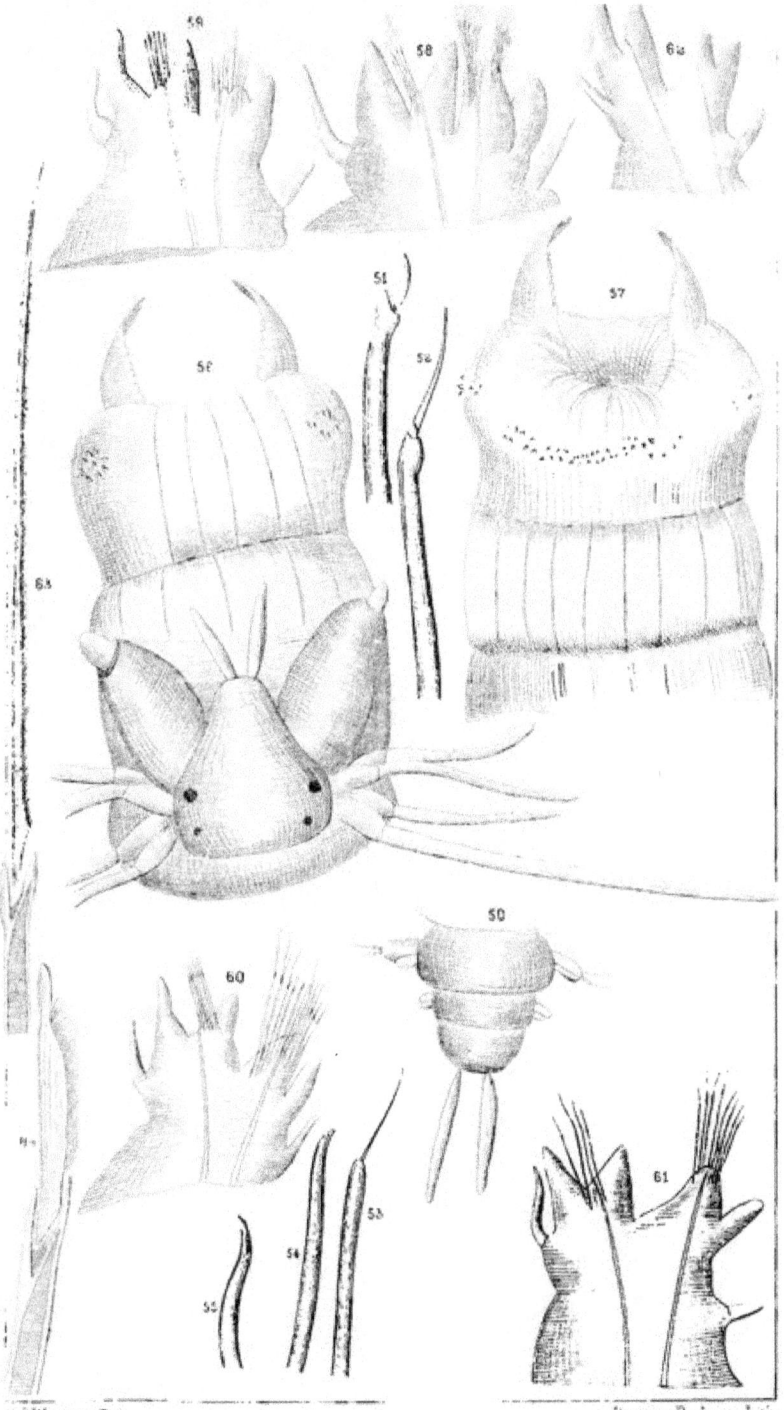

EXPLANATION OF PLATE VI.

NEREIS IRRITABILIS *n. sp.*

Page 231.

Fig. 65. Adult female, foot from 70th segment, × 20.
Fig. 66. Sexual seta, × 230.
Fig. 67. Adult male, foot from 70th segment, × 20.
Fig. 68. Adult male, 7th dorsal cirrus, × 20.
Fig. 69. Adult male, 8th dorsal cirrus, × 20.

NEREIS LIMBATA *Ehlers.*

Page 235.

Fig. 70. Asexual form, posterior segments, × 70.
Fig. 71. Sexual form, posterior segments, × 70.
Fig. 72. Female, foot from 10th segment, × 20.
Fig. 73. Female, foot from 20th segment, × 20.
Fig. 74. Female, foot from 50th segment, × 20.
Fig. 75. Female, foot from 70th segment, × 20.

MARPHYSA SANGUINEA *Quatr.*

Page 236.

Fig. 76. Compound bidentate seta, not found in adult, × 450.
Fig. 77. Simple seta, few in young, numerous in adult, × 450.
Fig. 78. Compound seta, few in young, many in adult, × 400.
Fig. 79. Acicula of upper bundle of setæ, × 450.
Fig. 80. Acicula of lower bundle, × 450.

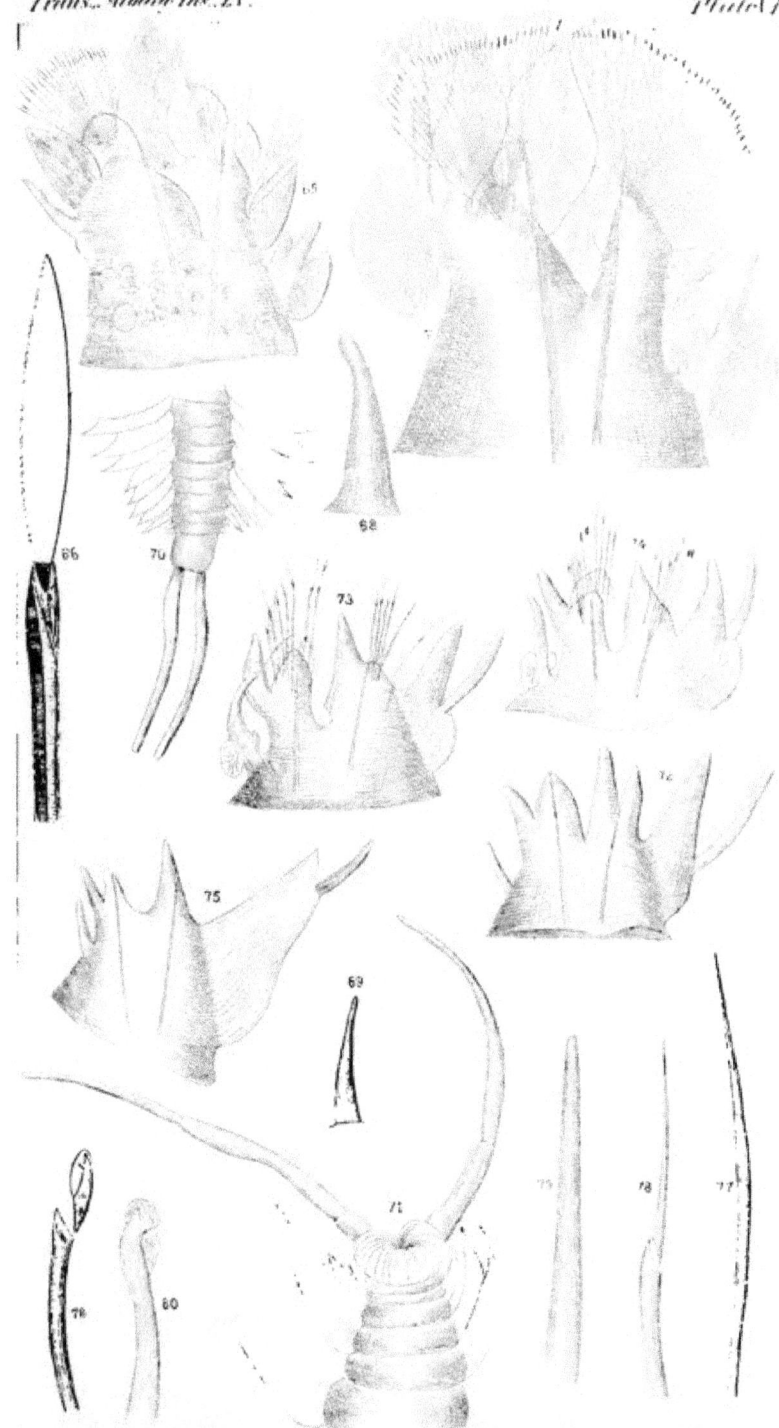

EXPLANATION OF PLATE VII.

MARPHYSA SANGUINEA *Quatr.*

Page 236.

Fig. 81. Series (c). Head and anterior segment, × 70.
Fig. 82. Series (d). Head with short lateral antennæ, × 25.
Fig. 83. Series (d). Head, × 35.

DRILONEREIS LONGA *n. sp.*

Page 240.

Fig. 84. Foot from middle of body, × 70.
Fig. 85. Outline of feet from posterior third of the body, × 70.
Fig. 86. Posterior foot, × 70.
Fig. 87. Setæ, × 230.
Fig. 88. Jaw pieces, enlarged.

STAUROCEPHALUS SOCIABILIS *n. sp.*

Page 243.

Figs. 89, 90, 91. Setæ, × 750.

TROPHONIA ARENOSA *n. sp.*

Page 245.

Fig. 92. Transverse section of 16th segment, × 15.
Fig. 93. Part of same segment, enlarged.
Fig. 94. Dorsal papilla, × 130.
Fig. 95. Dorsal seta from 16th segment, × 130.
Fig. 96. Ventral seta from 16th segment, × 130.
Fig. 97. Posterior ventral seta, × 130.

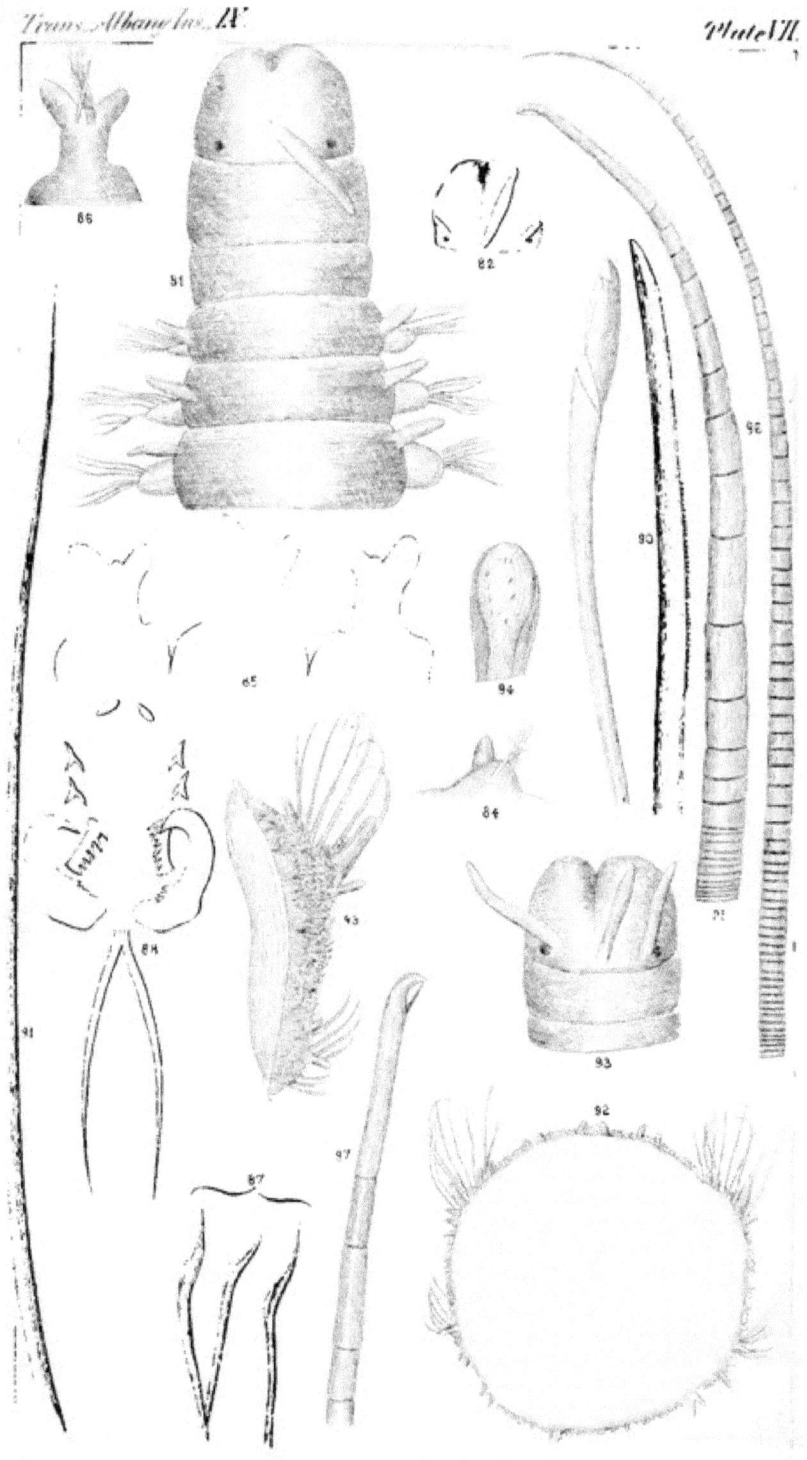

EXPLANATION OF PLATE VIII.

SPIOCHÆTOPTERUS OCULATUS n. sp.
Page 247.

Fig. 98. Head and buccal segment, × 70.
Fig. 99. Dorsal ramus from middle region, × 70.
Fig. 100. Dorsal ramus from posterior region, × 70.
Fig. 101, a, b, c, d. Setæ from anterior region, × 230.
Fig. 102. Peculiar seta of 4th setigerous segment, × 130.

NERINE HETEROPODA n. sp.
Page 249.

Fig. 103. Head and anterior segments, × 15.
Fig. 104. Foot from 7th segment, × 20.
Fig. 105. Foot from 21st segment, × 20.
Fig. 106. Foot from 25th segment, × 20.
Fig. 107. Foot from 35th segment, × 20.
Fig. 108. Seta, dorsal and anterior ventral, × 235.
Fig. 109. Ventral seta, behind the 25th segment, × 235.
Fig. 110. Seta with single margin, posterior ventral rami, × 235.

POLYDORA HAMATA n. sp.
Page 251.

Fig. 111. Capillary seta, dorsal, and anterior ventral rami, × 450.
Fig. 112. Setæ of 5th segment, upper series, × 450.
Fig. 113. Seta sometimes found in lower series 5th segment, × 450.
Fig. 114. Anterior ventral seta, × 450.
Fig. 115. Posterior ventral seta, × 450.
Fig. 116. Dorsal hook of posterior segments, × 230.

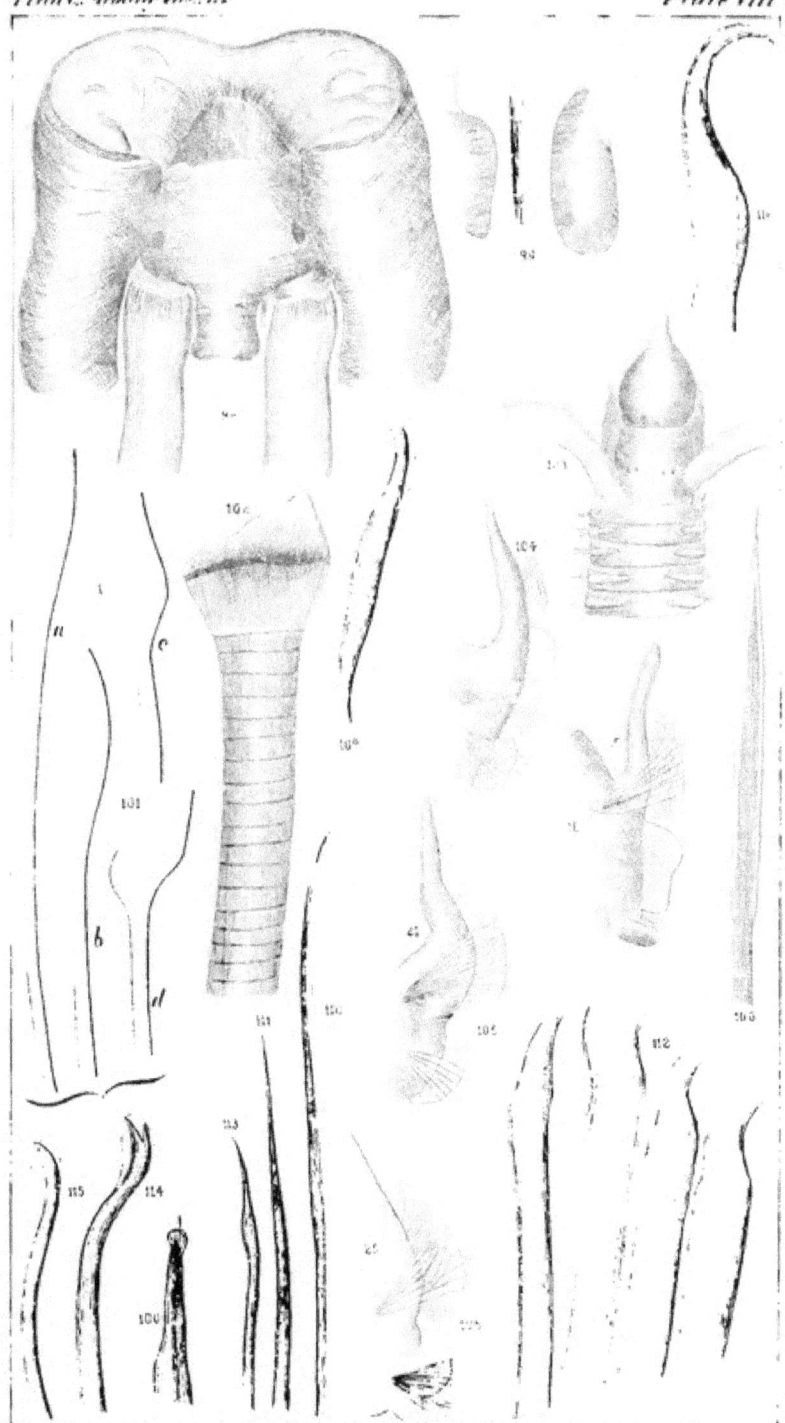

Plate VIII

EXPLANATION OF PLATE IX.

POLYDORA HAMATA n. sp.
Page 251.

Fig. 117. Long seta of posterior segments, × 450.
Fig. 118. Anal plates, × 25.

POLYDORA CÆCA n. sp.
Page 252.

Fig. 119. Long dorsal seta, × 450.
Fig. 120. Short dorsal seta, × 450.
Fig. 121. Setæ of 5th segment, × 230.
Fig. 122. Bidentate ventral seta, × 450.

ARICIA RUBRA n. sp.
Page 253.

Fig. 123. Anterior ventral setæ, × 450.
Fig. 124. Long ventral seta, × 150.
Fig. 125. Transverse section of anterior segment, × 20.
Fig. 126. Transverse section of posterior segment, × 20.

ARICIDEA FRAGILIS n. sp.
Page 255.

Fig. 127. Head and anterior segments, × 40.
Fig. 128. Transverse section of anterior segment, magnified.
Fig. 129. Transverse section of posterior branchiated segment, magnified.
Fig. 130. Transverse section of non-branchiated segment, magnified.
Fig. 131. Dorsal seta, × 450.
Fig. 132. a, b. Ventral setæ, × 450.

SABELLARIA VARIANS n. sp.
Page 259.

Fig. 133. Opercular seta of outer series, × 70.
Fig. 134. Opercular seta of middle series, × 70.
Fig. 135. Opercular seta of inner series, × 70.
Fig. 136. Dorsal seta of 3d segment, × 230.

Trans. Ent. Soc. IV. Plate IX.

EXPLANATION OF PLATE X.

SABELLARIA VARIANS n. sp.

Page 259.

Fig. 137. Dorsal seta of 4th and 5th segments. × 230.
Fig. 138. Dorsal uncinus, × 450.
Fig. 139. Ventral seta, outer half, × 450.

PECTINARIA (LAGIS) DUBIA n. sp.

Page 261.

Fig. 140. Paleola, × 70.
Fig. 141. Dorsal seta, outer half, × 450.
Fig. 142. Geniculate dorsal seta, outer third, × 450.
Fig. 143. Ventral uncinus, × 450.
Fig. 144. Spinule of the scapha, × 230.

MELINNA MACULATA n. sp.

Page 261.

Fig. 145. Uncinus, × 450.
Fig. 146. Capillary seta, × 230.
Fig. 147. Spinula, × 130.

LYSILLA ALBA n. sp.

Page 263.

Fig. 148. Seta, × 750.

POTAMILLA TORTUOSA n. sp.

Page 265.

Fig. 149. Long capillary seta from anterior segment, × 450.
Fig. 150. Short capillary seta from anterior segment, × 450.
Figs. 151, 152. Uncini from anterior segment, × 450.
Fig. 153. Capillary seta from posterior segment, × 450.

CABIRA INCERTA n. sp.

Page 267.

Fig. 154. Head and anterior segments, dorsal view, × 70.

EXPLANATION OF PLATE XI.

CABIRA INCERTA n. sp.
Page 267.

Fig. 155. Head and anterior segments, ventral view, × 70.
Fig. 156. Ventral hook, × 230.
Fig. 157. a, b. Dorsal setæ, × 450.

PHRONIA TARDIGRADA n. sp.
Page 268.

Fig. 158. Head and anterior segments, dorsal view, × 40.
Fig. 159. Foot from 6th segment, dorsal view, × 115.
Fig. 160. Foot from 6th segment, ventral view, × 115.
Fig. 161. Foot from 300th segment, dorsal view, × 115.
Fig. 162. Foot from 300th segment, ventral view, × 115.
Fig. 163. Transverse section taken at the 300th segment, × 20.

Trans. Albany Ins. IX.   Plate XI

H. E. Webster Del.

www.ingramcontent.com/pod-product-compliance
Lightning Source LLC
Chambersburg PA
CBHW031402160426
43196CB00007B/859